AF615425

The Mexican Caribbean

The Mexican Caribbean

Twenty Years of Underwater Exploration

Earl J. Wilson

Illustrated

Exposition Press Smithtown, New York

The stylized turtle is the trademark of CEDAM International with its world headquarters at Akumal, Quintana Roo. "Akumal" in Mayan means "Bay of the Turtle."

FIRST EDITION

Library of Congress Catalog Card Number: 81-69749

ISBN 0-682-49829-7

Printed in the United States of America

This book is dedicated to
my good friend,
Don Pablo Bush Romero,
a true pioneer
in the world of underwater
exploration

The Mexican Caribbean

1

Today, Robert Marx, outstanding diver and salvor, is one of the most knowledgeable people in the world about old shipwrecks, especially Spanish ones. However, in the latter 1950s he was just a penniless, young, ex-Marine Corps diving instructor eking out a living on Cozumel Island in the Mexican Caribbean, dreaming, diving, and searching for the treasure from an ancient shipwreck that would make him rich.

During the 1950s, skin and scuba diving were the fastest-growing sport in the United States. Many of this new breed had visions of treasure chests filled with gold to be found beneath the sea. Some, like Marx, were drifting down Mexico way, ready to take away whatever they might find. But Bob was no novice. As director of the Marine Diving School at Vieques, Puerto Rico, he had trained numerous leathernecks in the art of scuba diving. Actually, when I met him on Cozumel Island in 1959, the thought crossed my mind that he looked like an amphibian with his enormous chest and shrunken stomach.

Finally, he thought he had found his wreck. Mexican fishermen told him of an old Spanish shipwreck just off the jungled coast of Quintana Roo at a place called Punta Matanceros. It sounded promising.

The island of Cozumel had only a small fishing village. To get from there to the wreck site at Punta Matanceros one had to cross the sparkling waters by boat to the mainland coast. Here was the Territory of Quintana Roo. The jungle was everywhere—dense, exotic, almost impenetrable. It covered thousands of unexplored Mayan ruins and was filled with all manner of wild animals and colorful birds.

The area was sparsely populated. Some surviving Mayas lived a slash-and-burn subsistence in jungle clearings. Some picked chicle, a raw material for chewing gum, and others were fishermen. The crystal-clear waters offshore abounded in fish of every size. On land there were few roads; transportation was usually by boat along the coast.

For centuries Europeans had come to this area to colonize, loot, carry the faith, search for gold, war with one another. Some of their ships went down fighting; others were sunk by pirates or hurricanes.

In these early days, more often than not, efforts were made to recover valuable cargo from these shipwrecks using primitive means. Now and then fishermen tried with little success. Afterward these ships rested and rotted, undisturbed on the bottom, encrusted more and more by the coral, the wood to disappear, but the metal to lie waiting. Nautical archaeology was virtually unknown.

It was to this environment that Marx invited two of his diving friends, journalists, to come down from New York to Cozumel Island and help him explore the wreck at Punta Matanceros. They worked hard and found some interesting artifacts, buttons, beads, buckles, and such, but no gold. A story of their adventures appeared in the November 8, 1957, issue of the *Saturday Evening Post* and later in the book, *Diving for Pleasure and Treasure* by Clay Blair, Jr. They had barely scratched the surface of the wreck. They planned to return.

This was the beginning.

One of the first to don a scuba tank in Mexico was Pablo Bush Romero, or Paul Bush as he is called on the American side of the border. In the mid 1950s he had been the second president of Mexico's Frogmen's Club. Later, realizing that diving alone was not enough motivation to hold the club together, he proposed reforming the group into a club dedicated to underwater exploration, archaeology, and history.

Thus, early in 1958 a distinguished group got together in Mexico City to form CEDAM (Club of Exploration and Water Sports of Mexico), an acronym in Spanish for Club de Exploraciones y Deportes Acuaticos de Mexico. The organization would place "the sport of diving and the talents of its members at the service of country, science, and humanity." Pablo was elected executive president.

He, too, like Marx, was being drawn to Quintana Roo. Four years earlier he had met and married Elsy Betancourt of Merida on the northern side of the Yucatan peninsula. As an outdoorsman he had been interested when a friend told him of lakes in the southern part of Quintana Roo, "boiling with fish," and where jaguars would come down to the beach to eat the turtles. But most of all he was intrigued with a story told to him by his American friend, Emmett Gowen, expert fisherman, hunter, writer, and seafarer, who was on intimate terms with this part of Mexico's Caribbean coast.

Gowen had heard that inland, submerged under a lake called Boca Paila, was a "lost city of the Mayas." Late in 1958 Pablo organized a group on behalf of CEDAM to check this out. The group included others who would play a key role in CEDAM's explorations over the years to come: Chuck Ennis, Texas businessman, explorer and hunter; Victor Segovia, of Mexico's National Institute of Anthropology and History; Genaro Hurtado, underwater photographer; Alfonso Arnold, perhaps the best diver in Mexico; and Guillermo Calderón, motion picture producer.

After flying to Cozumel from Mexico City they crossed by boat and made camp on the mainland. Two launches and a native boat, outfitted with outboard motors they had brought, took them up canals constructed centuries ago by the Mayas. When they reached Boca Paila, no city—lost or otherwise—could be seen beneath the waves.

From here they proceeded up another canal to the Lake of Chunyaxche, finding at the mouth the ruined Temple of Xlabpak, then through another channel to Muyil Lake, where natives led them to a Mayan pyramid. Pablo was indignant. At both sites, despite their remote location, there was evidence that vandals had looted these archaeological sites. Even dynamite had been used in their search for treasure.

Mexico's National Institute of Anthropology and History (INHA) gave CEDAM credit for being the first organization to explore this archaeological zone using the old Mayan canals. But there was something more important. On the trip Gowen, by chance, had with him a copy of the *Saturday Evening Post* with the article about Marx, the Matanceros wreck, and Quintana Roo. He showed it to Pablo.

His diving activities already had alerted him—something not widely recognized at the time—to the danger of Mexico's losing historic artifacts from the sea to pillagers, foreign or national. CEDAM was dedicated to stopping this undersea looting.

Now Matanceros was a challenge.

In Mexico City Pablo took the article to the Mexican authorities. He pointed out that this wreck, in Mexican territorial waters, was soon to be revisited, without Mexican permission, by the American divers. He asked and was given clearance for CEDAM to explore this shipwreck "under the flag of Mexico" with the expedition directed by Mexicans.

When Bob Marx and his backers learned that CEDAM's first underwater expedition was to be held in the summer of 1959 to investigate the Matanceros shipwreck, they were outraged. They hired lawyers who claimed the "right of discovery." A behind-the-scenes power struggle began to develop, Americans against the Mexicans. That is when I became involved.

I was serving as deputy public affairs officer with the American Embassy in Mexico City and was assigned to look into the matter. Soon afterward I visited Pablo in his office. He was Mexico's leading distributor of Ford motor cars and trucks. He had shot big game all over the world. The Ford showroom was filled with numerous spectacular specimens of stuffed wild animals, including an enormous rearing polar bear, mixed in with the shiny new cars.

Pablo was in his fifties and looked years younger. He had a round face, a thin mustache, great liquid brown eyes, and was slightly balding. His English was perfect. Though a Mexican, and a nationalistic one, he loved the United States and understood very well the ways of Americans.

Later I was to learn he had been born in Mexico City in 1905, the son of an American father and a Mexican mother. To improve his English he had been sent to live with his grandmother in Chattanooga, Tennessee, where at twelve he became the first Eagle Scout in that city. At fourteen he returned to Mexico City, started as a grease monkey with Ford, and worked his way up to a position of wealth and influence.

From his boyhood, when he collected pottery, arrowheads, and other archaeological items, Pablo had been interested in that field and the outdoors. His hobbies included auto racing, big game hunting, deep sea fishing, skiing, wild animal training, and riding. Not long before I met him he had given up shooting game in favor of photographing the animals. He was a conservationist, had written three books on his adventures around the world, and lectured in many countries.

Over the years I was to find Pablo to be a good diplomat, a romanticist, a promoter, an adventurer, a keen businessman, affable, and warmhearted. We became good friends.

As to the upcoming expedition, Don Pablo was more than willing to have the Americans participate. They could even keep duplicate artifacts they might discover. The expedition would be under Mexican direction, but it could be a team effort. Though fuming at first, Marx and his friends accepted these conditions.

In that remarkable institution in Seville, Spain, the Archives of the Indies, where records are kept of Spanish voyages of trade and exploration during the Golden Age, research later turned up information on Matanceros. *Matanza* in Spanish means "the action of slaughtering," "massacre," "butchery," and it was possible the point off the beach of Quintana Roo gained this name after Mayas murdered the crew. Pablo and the others preferred to believe the name came from the fact the ship had been constructed at a shipyard at Matanzas, Cuba.

The ship's real name was *Nuestra Señora de los Milagros (Our Lady of the Miracles).* On November 30, 1741, the ill-fated ship left Cadiz, Spain, with 270 tons of mixed cargo bound for the New World. She went down on February 22, 1742, apparently sunk after a battle with a British ship, part of Admiral Vernon's blockading fleet. The admiral, a close friend of George Washington's brother, is how our first president's home, Mount Vernon, came by its name.

These and other details were not known when plans were laid for CEDAM's first expedition to begin on July 15, 1959, to probe Matancero's remains under the waves.

2

The CEDAM expeditionaries flew out of Mexico City on schedule. They were bound for Cozumel Island off the coast of Quintana Roo. There were four planes; three carried passengers, the other held cargo. For this first underwater expedition Mexico's national petroleum complex, PEMEX, loaned the group a C-82 freighter and a DC-3 executive plane. There were two small private planes, and later three navy helicopters would be added to the air arm.

The passenger planes landed at Cozumel and were met by military and government officials, the parish priest, and most of the villagers with their children. The cargo plane, with the expedition's supplies and equipment, for some reason was overdue.

A year earlier Mexico's Frogmen's Club had placed a statue of the Holy Virgin of Guadalupe under the sea at Acapulco as the New World's first underwater shrine. The Virgin had been proclaimed "Queen of the Seas," an honor confirmed by the Vatican. Now she was regarded as CEDAM's protector and her image was on their banner.

The group brought a small copy of this sculpture with them. The expedition began with the entire assemblage walking behind this image, led by a priest, to deliver it to the altar of the church where it and the mission would be blessed.

A local fishing schooner, *Cozumel,* had been hired to take the forward echelon of the expedition across the water to the mainland. Advance camp would be established under coconut palms lining the shore of a protected bay called Akumal, in the Mayan language meaning "bay of the turtle." This site was about two nautical miles up the coast from the shipwreck site at Punta Matanceros.

At four-thirty in the afternoon, members of the group were loading supplies they brought with them on the schooner. Word came by radio the missing cargo plane had crashed. It had gone down in the jungle at Champoton south of Campeche on the western side of the Yucatan peninsula. The crew of four, and Gregorio Castillo, a fisherman hired by CEDAM to look after the equipment, were all safe. It was Gregorio's first flight. They gave thanks that the Virgin had protected their colleagues.

Pablo sent three expeditionaries off immediately in one of the light planes to investigate the wreck and to salvage equipment. A police car picked up the trio at the Campeche airport. From there they sped to Champoton. A four-wheel-drive police vehicle took them up a mountain and through a pass to reach the wreck site.

At the scene it was evident all was lost. Both wings were torn from the C-82 plane. The fuselage was split down the middle. There was nothing to salvage among the twisted metal and smoldering ashes. It was indeed a miracle that those aboard had survived.

Back in Champoton the crew and the single passenger were still in shock. Nevertheless, they were able to talk. The pilot, Maj. Alfredo Bardaguer, told them sixteen minutes out of Campeche his port engine failed. Six minutes later his starboard engine stopped.

Castillo, the fisherman making his first flight, said when the trouble began the pilot turned and asked the others if they were Catholics. All said yes and they made the sign of the cross. The plane hit the trees. He didn't know how they all got out. Running from the wreck he heard it explode. Rescuers found them, he said, and took them to Champoton in the back of a truck.

In a report later on the expedition's lost equipment Pablo

described it as including: "A compressor, weighing 300 kilos, capable of filling a diving tank in little over one minute; a jeep; a decompression chamber; block and tackle; complete equipment for underwater lighting; several metal detectors for land and water; portable telephone; tents, hammocks; mosquito nets; axes, picks, blocks, cables; two 35-HP outboard motorboats; one underwater glider; several rubber boats; 25 sets of diving tanks; two sets of hookah equipment for sustained dives; a powerful radio for weather reports; medicine," and much, much more. A lesser man would have immediately scratched the expedition. Not Pablo.

On receiving the bad news he ordered part of the expeditionaries to proceed to Akumal to set up the advance base. They would have to do their best with the snorkels and swim goggles brought with them. Iron bars to pry the coral and a few other items were rounded up on Cozumel. For the time being, the group would have to provide its own food and water. Pablo would remain at the rear base on Cozumel to backstop their efforts and look for other supplies and equipment. To that end one of the small planes was dispatched to Mexico City with a CEDAM member instructed to search there and in Acapulco for an air compressor, an absolute necessity for diving tanks.

The crew of the downed plane was being flown back to Mexico City also. Castillo, the fisherman whose first flight had ended in disaster, when he heard Pablo planned to return him to Mexico City by transport other than air, declared he was a man and would return by plane with the others.

And so he did.

At Akumal the graceful palms of a coconut grove fringed the shore. The jungle was on one side, the powdery white sand of the half-moon beach on the other. The forward camp consisted of a few ponchos and sleeping bags. The fishing schooner floated easily on the transparent waters of the bay. For the first days the expeditionaries would have to live on coconut milk and fish.

First to go to the Punta Matanceros wreck site were the expedition's "point" divers, diving commander Alfonso Arnold and Sergio Castillo, and photographer Genaro Hurtado. They jumped into the water and began "lung" diving without scuba equipment. Almost immediately they located a huge anchor encrusted with coral. This was the mark—they were on target.

The ship, *Our Lady of the Miracles,* had run aground close to this rugged coast. The ship's wood had long since rotted. Looking down through the crystal-clear waters, there was no evidence of a ship, except for the anchor and several cannon. All one could see on the bottom were chunks of coral. The dive area went from close to the shore at three meters to five meters further out. Waves crossing the area exerted a heavy surge beneath the water.

Armed with iron crowbars, the divers attacked a block of coral near the anchor. Lifting it to the boat they were exultant to find encrusted on the bottom spoons, crucifixes, and colored beads, all held prisoner within the mass.

That night the divers sent their first message about these artifacts to HQ on Cozumel. It read: "Ship located, cannon and anchors visible. We recovered a few small pieces: glass, beads, spoons. All satisfied today's work. Tomorrow heavy work will begin. Speed us bars and compressors. Impossible to work without air."

Pablo already was on this. He had flown to nearby Isla Mujeres to meet José de Jesús Lima, who agreed to loan CEDAM his air compressor. Lima showed Pablo three old cannon he and his two sons had recovered scuba diving from a reef near Cancún. The relics were covered with coral, obviously very old. When the Matanceros expedition was concluded, Pablo agreed to send several CEDAM divers to examine this wreck site.

Meanwhile, the American contingent of the expedition arrived at Isla Mujeres, and Pablo met them. They were on the yacht *Pinch Hitter,* commanded by George Clark, president of the Yucatan Exploring Society. The valuable air compressor was loaded aboard *Pinch Hitter,* and they returned to Cozumel where it was transferred to a smaller boat and sent to the Akumal forward base camp.

At Cozumel the CEDAM president found a message had been received from the president of Mexico, Adolfo Lopez Mateos. It read: "I wish CEDAM great success in the first underwater exploration expedition conducted under the flag of Mexico by Mexican sportsmen."

This message was not entirely accurate because by now the American divers had joined the search. Aboard *Pinch Hitter* were Bob Marx, Clay Blair, Jr., author of the *Saturday Evening Post* article, and Wally Bennett, a *Time* photographer who had taken the pictures. Based on their earlier experience with the Matanceros wreck, they had devised a plan of attack which Pablo approved.

Scuba equipment now was available and the salvage work stepped up at once. The wreck site was marked into twenty sectors designated by buoys. Each sector was assigned to a diver who would be responsible for reporting areas on the bottom deserving more detailed search. Other divers would converge as needed.

It was exhausting physical labor. They were working in the surge with heavy hammers, chisels, and crowbars, fighting the undertow, trying to pry loose pieces of coral the size of a man's head or bigger. A diver would see something interesting only to be suddenly plucked away by the surge, often unable to relocate his find.

Each blow with a hammer on the coral brought clouds of colorful fish to feed on liberated organisms. Both above and below in the clear waters a careful watch was kept for barracuda and sharks.

Many of the divers suffered stomach problems, along with cuts and infections from the jagged coral. No one worked harder than Bob Marx. At one point, after staying on the bottom swinging a sledge and fighting the surge for more than three and a half hours, he came to the surface almost unconscious. The others thought it was the bends and sent him through stormy seas by small boat back to Cozumel. A Mexican navy doctor, trained in the U.S. for illnesses connected with diving, examined him. It turned out Marx was not suffering from the bends, but from malaria and exhaustion. He soon returned to join the others in the underwater search.

The divers found some 10,000 artifacts of 200 different types. The largest were four cannon "with signs of having been fired" and a one-ton anchor. One close call came when thirty divers, straining to hoist the first two cannon aboard their schooner, had the boat nearly driven ashore by a huge wave. Aside from these large objects, the rest of the artifacts were small, mined from the coral either on the bottom or from chunks taken back to the camp at Akumal or Cozumel.

Members of the expedition knew virtually nothing of the art of preservation of objects recovered from the sea. Few people did anywhere in the world. This science was not at all developed. They had to learn as they went along. Specialists and chemicals were brought to Cozumel for the arduous work of removing the coral and cleaning the artifacts. The divers were delighted to find that metal objects like bronze crucifixes, pewter spoons, and buckles regained their original appearance with only a little work with the polishing brush—this despite more than two centuries beneath the sea.

One striking discovery was made by Ramón Zapata. He found a ten-inch-long bronze cross. Later it was learned that this was a reproduction of a cross venerated in Caravaca (Murcia), Spain. It had an image of Christ on one side with the Virgin on the other. Thousands of smaller crucifixes were found, brought over for missionaries to use with the Mayas.

Another interesting discovery was a gold watch made in England. Inside, as packing to make a tighter fit for the case, was a small piece of newspaper. Incredibly, Bob Marx and Clay Blair, Jr., were able to trace this bit of paper back to a 1736 edition of the *London Daily Advertiser.* It was an ad for the sale of lumber—still perfectly intact.

The natives and fishermen of Quintana Roo had believed for more than 100 years that the Matanceros wreck held a treasure. They were sure gold was there, even if they couldn't find it. No matter who found it, the natives said the treasure belonged to them! In the earlier exploration Bob Marx and Clay Blair, Jr., secretly buried some of their artifacts on the beach. They made a map and left them there because the natives thought the Americans had found gold and were increasingly hostile.

The locals kept watch on CEDAM efforts, and when head diver Alfonso Arnold came up with a packet of metal sheets whooping he had found gold, this did not go unobserved. Actually, the sheets were copper. That night a group of locals armed with machetes came to the divers' camp at Akumal. To prevent any harassment the Mexican navy sent a sloop to Akumal to lie offshore. Some marines were detailed to guard the camp.

As far as anyone knows, there was never any gold on Matanceros. It was a Spanish law in those days to have at least two divers on each ship to salvage in case of shipwreck. Research in the Archives of the Indies revealed that after *Nuestra Señora de los Milagros* went down, divers carried out a salvage operation. The recovered merchandise was taken to Campeche and Veracruz. Some of this was silk cloth from England. This was regarded by the Spanish as contraband, sufficient for a death sentence. But, according to the old records, the Marquis of the Casa Madrid denied any knowledge of such cargo carried on his ship.

Matanceros was a seagoing store loaded with merchandise. *National Geographic* shows the location of the wreck on its 1977 map entitled "Colonization and Trade in the New World." Because of the varied wares it called Matanceros the "Five and Dime" wreck.

Altogether the divers found a miscellany of merchandise from five countries: Spain, Italy, Germany, England, and France. These artifacts were unique and of special interest to archaeologists because they were on a ship with cargo coming to the New World, rather than leaving it, which was more often the case with old shipwrecks of this era.

Aboard were all sorts of things for the colonists: packets of needles from Aachen, Germany; hundreds of religious medals; thousands of crosses; Spanish coins; spectacles; beads; buckles; spoons; cannonballs; handles of tools; thimbles; glass cups; plates; bottles; and pewterware.

The success of this mission brought visitors, and I was one of them. I flew from Mexico City to Cozumel, where I boarded a Mexican navy sloop. Aboard were journalists representing the *London Times, The Observer,* Brazil's *O Cruzeiro, Paris Match,* and others. I was happy to find my friend, Alma Reed, an American journalist living in Mexico City and CEDAM's historian, also a member of the group.

We dropped anchor off Punta Matanceros. I could see the anchor on the bottom through the clean, clear water, snagged on a sea bottom littered with chunks of coral. On the shore the beach was mainly jagged coral with sand patches here and there. Behind that was the jungle. Overhead arched the beautiful blue tropical sky. Gulls circled. We could see some divers working beneath us. I changed to bathing trunks, donned a snorkel, and jumped over the side. As I swam downward I could feel the surge of the waves passing overhead, and I understood the difficulties experienced by the divers. As I drew near to one of them he pointed to a piece of coral, signaling for me to pick it up. It was about the size of a football. I carried it to the surface, put it on the boat, and examined it. Embedded on the underside were several bronze crucifixes, entombed in the coral for more than two centuries! I was able to pry one loose. Pablo insisted I keep it. It still hangs from a chain around my neck.

Other visitors arrived in the closing days of the expedition to congratulate the CEDAM group. The chief of the Seventh Naval Zone, also a CEDAM counselor, came by helicopter from Isla

Mujeres. Another chopper brought the secretary of national defense, Gen. Augustin Olachea, and the governor of the Territory of Quintana Roo, Sr. Aarón Merino Fernández.

CEDAM president Pablo Bush came over from Cozumel by helicopter for the final breaking camp ceremony at Akumal. The divers stood at attention while the flags of Mexico, the United States, and the standard of Our Lady of Guadalupe were lowered in a solemn ceremony.

This accomplished, Pablo flew off to Cozumel, leaving the twenty remaining expeditionaries to come over from the forward camp by boat with their gear and "goodies." At Cozumel, after Pablo alighted, the chopper took off again and tried to land by the main pier where suddenly the left pontoon struck coral rocks and exploded. The plane hit the water, upended, and the whirling blades chopped up the aircraft. The pilot, the crew, and three passengers escaped unhurt.

From their boat the divers returning from Akumal at sunset could see from the distance the wreck of the helicopter. It had been recovered from the water and was lying on the dock like a broken bird. It was surrounded by a curious crowd.

Pablo stood on the dock shouting to the divers across the water, "It's nothing! We're all alive!" The Virgin had protected them again.

Mendel Peterson, curator of the Armed Forces History Department, and later director of the Underwater Exploration Project, both of the Smithsonian Institution in Washington, D.C., had been invited down to examine the artifacts. Peterson himself was a pioneer at underwater exploration, an authority in such matters.

Because of the quantity and diversity of the salvaged objects, Peterson hailed the Matanceros as the richest shipwreck yet explored in the Western Hemisphere.

Nautical archaeology was in its infancy. The CEDAM group knew they were novices. Nevertheless, they were elated with the results of their first underwater expedition. Alma Reed, representing CEDAM the following year in Barcelona, Spain, reported this "birth of Mexican submarine archaeology" to the

100 delegates gathered from many nations for the Third International Congress on Submarine Archaeology.

Alma, writing from Barcelona, where Columbus had reported to Ferdinand and Isabella his amazing discoveries from his first voyage, said it was fitting that this same city should serve as an open forum now "for the consideration of an equally vast enterprise—nothing less than the exploration of another New World—the immense extension of ocean and seas that occupies almost three-quarters of the surface of the planet."

Matanceros had whetted the appetite of the CEDAM divers. They were eager to go forward to other explorations under the waters of Quintana Roo.

3

While the CEDAM divers were excavating the Matanceros wreck, Pablo Bush, their president, was having other wreck sites scouted for future expeditions. To that end, photographer Genaro Hurtado and several other divers were sent up the coast by boat to a place called Punta Cadena.

A fisherman said there was evidence of a wreck just off the beach. It had taken place earlier in the century. Local legend said this ship, a Chilean vessel with a cargo of European wines, went aground in a storm. The crew fled as natives boarded the ship, smashed open the casks with their machetes, and downed the whiskey, cognac, and wines from hastily made coconut shell cups. They said this bash went on for six days and, as a result, more than fifty of the natives died.

Hurtado and the others found the coast at Punta Cadena to be rugged, composed of jagged coral, and washed with large waves. *Cadena,* in Spanish, means "chain." Fighting the surf, the divers did, indeed, sight an enormous anchor chain on the bottom encrusted with coral. Otherwise they found little of value, only some ship's nails and a few pieces of bronze.

The Punta Cadena wreck was logged and eliminated as being of no further interest.

Keeping his promise, Pablo returned to Isla Mujeres to see José de Jesús Lima and check out the wreck site he and his two sons had discovered. By now famous American inventor Ed Link, famed for developing the Link Trainer, a flight simulator used for teaching pilots, had arrived at Cozumel at Pablo's invitation in his well-equipped salvage boat, *Sea Diver II*. Link was one of the nation's leaders in the development of devices for undersea exploration. Pablo and several CEDAM divers went on Link's boat to explore the Cancún wreck site. The Mexican navy corvette *Baranda* accompanied them.

Luck was with them. On the very first dive three cannon were located. These were recovered and hauled back to Cozumel. Mendel Peterson, the Smithsonian expert, examined them. He concluded the ship was probably Spanish and had gone down sometime before 1550. This would make it the oldest wreck yet discovered in the Western Hemisphere. The cannon might have been made before the discovery of America.

The CEDAM group speculated the ship actually was *La Nicolasa,* flagship of Francisco de Montejo, conqueror of the Mayas, whose ship went down in the early 1520s.

Pablo decided this and other wreck sites in the area would be explored during CEDAM's second expedition the coming year.

Before leaving Cozumel, some of the Matanceros artifacts were left to establish a small museum. It was Pablo's expressed hope that, aside from being a tourist attraction, the display would encourage natives to turn over relics they might find to the proper authorities, rather than sell them to tourists.

After being cleaned and catalogued, the best of the Matanceros artifacts, released by the Mexican government, were used to establish a small CEDAM museum in Mexico City. These museums, though unpretentious, were believed to be the first in Latin America, if not the world, for the exhibition of objects recovered from old shipwrecks.

Isla Mujeres, "Island of Women," derived its name from the number of Mayan idols representing women found on it by European visitors in earlier times. In December 1959, Pablo,

Raúl Gómez, and Alfonso Arnold set out from Mexico City for Isla Mujeres to make a preliminary reconnaissance of it as a possible base for the upcoming summer 1960 expedition.

Pablo and Gómez were in a car towing a small outboard motorboat. Arnold drove a truck loaded with diving equipment and other supplies. Approaching Jalapa a wild driver, probably drunk, came head on at Arnold's truck at high speed. He spun the wheel and swung off the road. The speeding car barely missed the other CEDAM car and disappeared down the road. Arnold's truck somersaulted and came to rest upside down. The diving gear and other supplies were scattered around the area, but Arnold was not hurt and the equipment was undamaged. They gave thanks that once again the Virgin had extended her protection.

A pickup truck stopped by the wreck. Two Mexican youths came up to the CEDAM group and offered to load the scattered equipment into their vehicle. They would take it wherever the CEDAM group was going. This was not a commercial proposition. They simply wanted to help. Pablo said their destination was Puerto Júarez at the northeastern end of the Yucatan peninsula. The youths didn't know where this was but told him to lead on. Their tires were thin, so Pablo told them to take the good tires from Arnold's wrecked truck. CEDAM had acquired two new members.

Driving this route through Mexico in 1959 was in itself a minor adventure. They drove to the port city of Veracruz on the Gulf of Mexico, then down the coast to San Andres Tuxtla where they passed the night. Continuing along the narrow road they took ferries to cross the many rivers, the Coatzacoalcos, the Tonalá, the Carrizal, Paso de los Muertos, Usumacinta, Frontera, and the San Pedro. The second night they reached El Zacatal in Laguna de Términos, just missing the ferry across to Isla del Carmen.

At dawn they crossed on the ferry, bought precious gasoline, and drove to Puerto Real at the island's end. Here they boarded another small, rickety ferry which they shared with a truck loaded with decaying fruit. They reached Campeche on the eastern side of the Yucatan peninsula that night.

On the fifth morning, after overhauling the trucks and buying provisions, they headed for Mérida, capital of Yucatan. Here American fisherman-hunter-writer Emmett Gowen joined the group. The last push was across to Puerto Juárez, not much of a port, just a few huts on the beach opposite Isla Mujeres. There the group put their outboard motorboat in the water and crossed over to Isla Mujeres, their final destination, where the Mexican navy had a small base.

The Mexican navy continued to give CEDAM its full support. The secretary of the navy had written to Rear Admiral Armando Cañizares, commanding the Seventh Naval Zone from his headquarters on Isla Mujeres, to extend all assistance. The admiral greeted the visitors and sent a navy launch back to Puerto Juárez to pick up their diving equipment and the rest of their gear. Their air compressor for the scuba tanks had been damaged in the accident. This was repaired in the navy's machine shop.

Besides this help there was a piece of good luck. The admiral invited them to accompany him the next day to Cozumel Island. They would take part in a banquet and other festivities in honor of Mexico's president, Adolfo Lopez Mateos. He was making a swing around this part of the country. They accepted the invitation and Pablo had the chance, if only briefly, to focus the president's interest on the underwater explorations CEDAM had made in the Mexican Caribbean.

There was also bad luck. The wind was high and the sea was rough, cutting short the survey at sea. The group went by boat as far north as Isla Contoy. They also dove briefly on the Cancún wreck, but the weather was against them.

On land they explored Isla Mujeres, once headquarters for pirate Jean Lafitte. Aside from the navy facilities there wasn't much. Most numerous among the inhabitants were the *garrapatas,* ticks.

Despite the drawbacks it was determined Isla Mujeres would be the base for the first part of CEDAM's second underwater expedition in the coming summer.

Most of the *expedicionarios* arrived in mid-June 1960, as planned, at Puerto Juárez with their equipment. There was a lack of boats. The divers straggled across to Isla Mujeres in a heavy rain, and the last arrived close to midnight. Few rooms were available and these were far apart. There was no transportation. If they didn't know it before, the divers knew it now—they were in the boondocks.

In the morning they assembled in the small town to form a procession behind an image of the Virgin of Guadalupe, Queen of the Sea, winding their way to the church. Then, at the naval base, in the presence of the governor of the Territory of Quintana Roo and other military and civil officials, the CEDAM flag was raised. The expedition officially had begun.

That evening a banquet was held for the American and Mexican participants. There were fireworks, music, and Mayan dances. Gideon "Gid" Neal, head of the Caribbean Archaeological and Exploring Society, presented a letter from the governor of Texas to the governor of Quintana Roo. William de Mello, representing the Middle American Archaeological Society, presented a letter from the governor of California. Glasses were raised, toasts were offered, and spirits lightened.

The first task of the expedition was to anchor a large raft to serve as a diving platform over *La Nicolasa,* the Cancún wreck. There was a stiff wind and high waves, and this activity took two days. Nearly two tons of heavy chain and a great deal of thick cable were used to secure the raft to the coral reef.

Working from this platform, the divers located yet another cannon on the Cancún wreck. It was concreted in the coral. With the weather worsening they put off recovery. Work came to a complete halt when waves set the raft adrift. It was blown to sea. A navy patrol boat towed it back.

Meanwhile another dive team had gone north to explore a wreck they called *El Dormitorio*. They found the site but located few artifacts. Rumor said this ship was about 100 to 150 years old. It would date roughly from the time pirate Jean Lafitte had roamed the waters around Isla Mujeres.

With the foul weather Pablo was finding it difficult to control the separated dive teams. The expedition had gotten off to a poor start. At base camp housing and transportation were bad. As the days went by, half the expeditionaries were on the sick list with little accomplished. The first week was judged almost a total loss.

In the second week the weather cleared. The divers found the Cancún cannon again and raised it. Gid Neal had returned to the States, leaving behind an underwater metal detector. Using this, CEDAM divers located a large anchor, broken in two parts, at the same wreck site. Under it, when it was raised, they found tile and pieces of ceramic.

Expanding their underwater survey of wreck sites, one dive team went in the navy patrol boat, *Pez Vela,* to explore a site at Isla Blanca in the north. They found some bronze nails, chains, and two empty metal boxes. They named the site after Lafitte.

Moving on toward the reefs of Islache, one of the CEDAM divers, Sergio Castillo, jumped over the side before the boat anchored. He immediately popped back to the surface yelling: "Cannon! There's a cannon here!"

This wreck site had been suggested by Admiral Cañizares. As a result, they called it the Cañizares wreck. A telecompass was used for fixing the precise position of this wreck. This device had been brought along by John Ferris, president of the Middle America Archaeological Society.

The small cannon Castillo had located was imprisoned in an enormous block of coral. The entire chunk was raised and taken back to Isla Mujeres. When they freed it of the coral, they found the weapon had its load of powder and fuse intact. A wooden stopper was in the muzzle to protect the powder from dampness. When the cannon went to the bottom of the sea, it had been ready for firing.

The divers mounted the piece on an old navy tripod and placed it, pointed out to sea, in front of CEDAM's headquarters on Isla Mujeres.

An aerial view of Boca Paila on the Quintana Roo coast, entrance from the sea to a network of freshwater lakes and old Mayan canals used by CEDAM to explore the area. (Photo by Genaro Hurtado)

CEDAM explorers in 1958 found this Mayan pyramid at Muyil had been looted. Inside they found a nest of deadly, poisonous snakes. (Photo by Genaro Hurtado)

An aerial view of "the castle" in the Mayan walled city of Tulum on the coast of Quintana Roo. In the late 1950s, when CEDAM first visited this ancient city, it had been seen by few outsiders. (Photo by Genaro Hurtado)

On CEDAM's 1959 expedition their cargo plane crashed in the jungle en route from Mexico City to Cozumel. No one was hurt, but the plane and supplies were a total loss. (Photo by Genaro Hurtado)

An aerial view of Akumal (Turtle Bay) and the coconut plantation fringing the shore in front of the jungle where in 1959 CEDAM divers made their advance base for excavating the Matanceros shipwreck. (Photo by Genaro Hurtado)

Dressed in long underwear as protection against coral, armed with a chisel and sledge hammer, a CEDAM diver jumps into the water at the site of the Matanceros shipwreck. An air hose (hooka) connects him with the ship's air compressor, allowing longer work periods underwater. (Photo by Genaro Hurtado)

Divers search for artifacts from a wreck believed to be *La Nicolasa,* flagship of the fleet of Montejo, conqueror of the Mayas. (Photo by Genaro Hurtado)

American and Mexican CEDAM expeditionaries en route by fishing boat from Akumal to Cozumel Island after five weeks of diving and exploration of old shipwrecks off the Quintana Roo coast. (Photo by Genaro Hurtado)

Entrance to cave at Xelhá lagoon, explored in 1960 by CEDAM divers. Swimming inside they found an altar and Mayan artifacts. (Photo by Genaro Hurtado)

CEDAM erected this tombstone over the grave of pirate Jean Lafitte. Descendants of Lafitte, town officials, the American consul general, and others joined CEDAM in the dedication ceremony. (Photo by Genaro Hurtado)

Two ancient cannon recovered in 1960 off Cancún are examined at Isla Mujeres by CEDAM founder Don Pablo Bush Romero *(right)* and Vice-Admiral Don Armando Cañizares of the Mexican navy. (Photo by Genaro Hurtado)

CEDAM divers prepare airlift to raise artifacts from the Sacred Well of Chichén-Itzá in 1961. Ancient Mayas were believed to have sacrificed virgins in the well to appease their rain god, Chac, and also to have thrown in valuable ornaments. (Photo by Genaro Hurtado)

"Jungle Maya" guards kept a constant watch on a temple said to contain their "Book of God." In 1962 CEDAM visitors were not permitted inside the temple. (Photo by Genaro Hurtado)

Sick and near death, Juan Bautista Vega *(left)*, held prisoner since his youth by "Jungle Mayas," was rescued in 1962 by a CEDAM expedition. A suspicious Mayan priest *(right)* eyes the visitors from the outside world. (Photo by Genaro Hurtado)

Joe Kelly Hughes *(left)*, former U.S. Navy Seal, and Genaro Hurtado, Mexican underwater photographer, examine artifacts recovered from a shipwreck on Scorpion Reef in 1968. (Photo by Dr. Andreas Rechnitzer)

CEDAM divers prepare to enter waters of the Lake of the Sun in the Volcano of Toluca in search of relics from the past. Special diving tables had to be made to check water pressures at this high altitude. (Photo by Pablo Bush)

This first phase of the expedition had been directed toward the north. Now it was time to turn south. Headquarters was shifted from Isla Mujeres to Cozumel Island.

Advance base was set up on July 4, 1960. Mexican and American flags were raised at the chicle outpost of Tancah, on the coast of Quintana Roo, just north of the Mayan walled city of Tulum.

With stormy weather and high surf making diving at sea from the boats difficult, Pablo improvised an exploration of a nearby *caleta* or natural cove on the coast which had been cut into the soft limestone by the action of the waves. It was called Xelhá. Nothing was known about it, and they had no maps. They thought it quite probably was a Mayan port in the old days.

The party was divided into two groups. Pablo with one group would go by boat, approaching from the sea. The leading diver, Alfonso Arnold, with the others would go by land, using a Jeep to push through the underbrush.

During the day's exploration, no contact was made between the two groups though they shouted on either side of the jungle growth that separated them. That night, back at the Tancah camp, each group had something of interest to report.

Pablo's team had discovered what they believed to be a fortification built by Francisco de Montejo in his conquest of the Mayas in the early sixteenth century. It was a wall, a crude rampart of stones. The stones had been removed from nearby Mayan monuments. The wall was built across the small peninsula. It was said Montejo, pushing on against the Mayas, had left a garrison of forty men on the peninsula with the wall as protection on the land side. They also found foundations of the soldiers' huts. Some historians say when Montejo returned four months later, the soldiers he had left were gone, leaving not a trace.

Arnold reported his group, swimming in the inlet, had discovered an underwater cavern. Shoulder-deep, they waded through the water at the mouth of the cavern. It had a low ceiling, and inside it widened. There was a small island in the center. In the roof overhead were two orifices giving light inside the chamber. On the island was an altar which they thought was possibly a Mayan crematorium. They picked up some potsherds, two small jade axes, and a ceremonial grinding stone.

The next day, accompanied by Braulio Garcia, of the National Institute of Anthropology and History (INHA), the whole group returned to the cavern for further exploration. More potsherds were discovered. They were judged by the expert to be from the Maya-Toltec Era, the first such to be "officially" discovered in Quintana Roo. It was clear from traces of destruction on the altar that "pirates" had been there before them.

Despite continuing bad weather, the CEDAM group again divided to continue their work at sea. Arnold led one group of divers back to the Matanceros wreck where they dynamited the stubborn coral for the first time. After the explosion, they recovered many more small artifacts. All were more or less duplicates of those recovered in earlier operations.

The other group, operating from the motor-sailboat *Pez Vela* and the navy patrol boat *Azueta,* went to investigate a site south of Tulum. They had been tipped to this by Argimiro Argüelles, captain of the schooner *Cozumel.* Diving on the site, the CEDAM expeditionaries discovered seven cannon, an anchor, and some grapeshot. They named this the Argüelles wreck.

Meanwhile, Pablo had returned to base camp on Cozumel Island. There he was notified by the navy of the approach of Hurricane Abby. They were ordering their patrol boat, *Azueta,* to return to the navy base on Isla Mujeres for protection. It was time to terminate the expedition. The problem was neither of the two boats had a radio. There was one fast way to communicate with them.

A small navy plane took off from the runway at Cozumel

and succeeded in locating the motor-sailboat *Pez Vela.* It was at anchor near CEDAM's forward camp at Tancah. The pilot dropped a bottle containing a message Pablo had written telling them of the danger and to return at once. The pilot was unable to locate the navy boat.

By coincidence the American naval attaché from Mexico City, Capt. Jake Heimark, had landed at Cozumel in his DC-3 transport plane. Jake was on an official trip to Belize in British Honduras. Learning of the difficulty, he offered to fly Pablo to Tulum, on the way to Belize, to see if he could locate the navy patrol boat from there.

The chief of the Mexican Air Command on Cozumel pointed out to Jake the rocky runway at Tulum was short and completely abandoned. Only single-engine planes had ever landed there. Despite this warning and the oncoming hurricane, the American attaché plane took off from Cozumel. In a few minutes it circled low over CEDAM's forward camp at Tancah, then landed with precision on the rocky runway at Tulum, the ancient Mayan walled city on the seacoast. Seeing the plane low overhead, several of the CEDAM divers jumped in their Jeep and drove to the Tulum airstrip. They told Pablo they had been in contact with the *Azueta.* The whole group and both boats were set to return to Cozumel.

From Cozumel, where Pablo and others were picked up, the expedition members raced in front of the hurricane to Isla Mujeres. Then they went over storm-whipped waters to Puerto Juárez on the mainland, then driving through high winds and stinging rain to Valladolid. Here Pablo, with several others, planned to investigate several *cenote*s, or limestone sinkholes so common in Yucatan.

When the storm abated they visited several *cenote*s. Photographer Genaro Hurtado, with only a snorkel and weights, jumped into one of them. He hit bottom at sixty feet. It was rocky down there. Lights and proper diving equipment would be needed for any close examination.

The CEDAM president had become fascinated with the

possibilities of *cenote* diving. He believed some of these undoubtedly held secrets from the past. He was particularly interested in the famous Mayan Sacred Well of Sacrifice at Chichén-Itzá. When they reached this *cenote,* he studied it carefully. By now he had various possibilities in his notebook for CEDAM's third expedition. The Sacred Well was especially intriguing. He decided it would be a future target.

4

The ghost of pirate Jean Lafitte seemed to hover over CEDAM's 1960 expedition. Lafitte, of course, earned his place in history when he marshaled his pirate band to aid Gen. Andrew Jackson at the Battle of New Orleans against the British in 1814. Though pardoned for past crimes for this patriotic act, Lafitte again returned to piracy, burned Galveston in the 1820s, established himself on Isla Mujeres in 1821, and became for a time a scourge to shipping in the region.

Earlier in the year, when Pablo and his companions had set out to reconnoiter Isla Mujeres in preparation for the upcoming expedition, he had stopped in Merida, capital of Yucatan. There Lafitte's name came up in conversation with a friend, Luis González. Luis said he had heard the pirate was buried in a tiny fishing village called Dzilam-Bravo on the north coast of Yucatan.

Pablo's interest was piqued. This sounded like a worthwhile add-on investigation. Consequently, just before the start of the second expedition, he met in Merida with a small group of Mexican and American expeditionaries to begin the search for the pirate's grave and to establish contact with his descendants.

Traveling to Dzilam-Bravo, the group's first inquiries were made to several village inhabitants sunning themselves in the little plaza. One man led them to the home of Don José M. Estrada, a white-haired old gentleman who confirmed that his family was indeed descended from Lafitte. He presented his family, blond and blue-eyed, bearing no resemblance to the natives of the region. Genealogical information was provided.

The legend was, Don José told them, that Lafitte had come to Dzilam-Bravo with a beautiful mulatto lady, Lucia Allen, he had abducted from Mobile. He had wanted to live a quiet and honest life, hiding from the world in this remote village. There he died and was buried. Later a daughter was born.

The old gentleman conducted the CEDAM search group to the village's "new" cemetery, established when the sea began devouring the old graveyard through erosion. They were shown a wooden, worm-eaten plaque at the head of one of the graves. Carved on the slab was a rose, a cross, and the words: "Jean Lafitte, Re-exhumado, 1938."

It was enough. They were convinced. Here rested the remains of Lafitte the pirate. Pablo offered Don José a deal. In return for the wooden marker from Lafitte's grave, which would go into the CEDAM museum, CEDAM would provide a new marble plaque. Don José agreed.

Thus it was, a month later, when the expedition's tasks of diving on the old wrecks and the jungle exploration had been completed, that the same little CEDAM group again assembled at Dzilam-Bravo's small cemetery. True to his word, Pablo had the new marble plaque. On one side was a reproduction of the words on the original wooden marker along with the rose and the cross. On the reverse side was a list of the names of the CEDAM members who had participated in this small adventure.

It was a big event. Most of the townspeople had gathered at the gravesite. A priest officiated. Among the journalists was Alma Reed, representing several English-language papers in Mexico City. She was also there in her capacity as CEDAM historian.

The story appeared in the Mexico City press and an item was carried by the wire services. The report provoked an unexpected controversy waged in letters to the editor. Especially outspoken was J. Ignacio Rubio Mane, author of a 1938 book entitled *The Lafitte Pirates*. He claimed the Dzilam-Bravo tomb was that of Lafitte's brother, Pierre. Jean Lafitte, he said, had died in a naval battle while fighting in the service of Latin American patriot Simon Bolivar.

Others questioned CEDAM's claim to have located Lafitte's grave. A reporter for the *Times Picayune* in New Orleans asked: "What of those people who believe that Lafitte is buried in the Bay de Barataria?" The reporter wrote that though Lafitte was most certainly dead, his story remained very much alive.

In New York's Central Library, Alma Reed did some historical digging. There she unearthed the "diary" of Lafitte, which had been published by the pirate's great grandson, John Laffite, who had Americanized his surname. He wrote it had been the buccaneer's wish that the diary not be published until 107 years after his death.

Out of all this one thing became clear: when the Lafitte brothers ran into difficulties one convenient out was to "die" and be "buried." Later they could return, as if by magic. Because of this Pablo wrote: "Jean Lafitte is buried in Dzilam-Bravo, Yucatan; in Guanajay, Cuba; in the waters of the Atlantic, in Barantaria . . ."

But Alma Reed, still tracking, talked with the pirate's descendant who told her that John Laflin, last alias of the gentleman pirate, was the head of an honorable family. He died in Alton, Illinois, on May 5, 1854, and was buried there in the family cemetery.

There, he said, rest the remains of Jean Lafitte.

Or do they?

5

Submarine archaeology, or nautical archaeology as it is now called, first began in the Western Hemisphere at the Sacred Well of Chichén-Itzá, the fabulous Mayan city on the Yucatan peninsula, at the beginning of the twentieth century. In early 1961 CEDAM for its third expedition set out to excavate the depths of this historic *cenote.*

This new investigation brought to mind the names of three persons associated in the past with unraveling mysteries of the Sacred Well. Only one, Alma Reed, CEDAM's historian, was still alive. The others were Edward Thompson, former American consul general in Yucatan, and Felipe Carrillo Puerto, former governor of the region. Fate had drawn these three together.

Thompson was an avid amateur archaeologist. At the turn of the century, in this remote part of the world, he had purchased for $75 the area of what is now the Chichén-Itzá archaeological zone. He assiduously studied the ruins, learned the Mayan language, took a Mayan woman, fathered her children, and obtained the post of American consul general through sponsorship of the Antiquarian Society of Boston.

What came to intrigue Thompson most was the legend of the Sacred Well. It was said in ancient times, when there was pestilence or a famine, the Itzás practiced a cruel ritual to appease the gods.

The ceremony began in a small temple on top of the main

pyramid known as *El Castillo.* From this edifice, escorted by priests wearing black masks, would emerge a young maiden dressed for her marriage to Chac, the water god who lived in the depths of the nearby Sacred Well.

Wearing a wreath of white roses, the maiden was conducted toward the well to the sound of shrill notes from reed whistles and the steady beat of the death drum. At the *cenote* the music increased in intensity as the priests and others lined the rim of the well, crying and chanting, tossing in pots of smoking incense, and in their ecstasy gold ornaments and other jewelry.

The maiden, who had been drugged, wore a copper bell suspended from a chain around her neck. As a last act the head priest would step forward and tear the clapper from this bell as the sign of death. The maiden was then thrown into the water. Her marriage to Chac would save her people from disaster.

Was the legend true? It haunted Thompson. He decided to find out and began diving into the well late at night to conceal his search from the natives. He often came close to death, but he did succeed in bringing to the surface human bones of a girl between twelve and sixteen years of age.

Thompson redoubled his efforts, brought in a Greek hardhat diving master, and learned diving skills from him. In 1903 a wealthy American provided him with funds for a simple dredge operated by manpower. Using this a great many more human bones, mainly those of young girls, were recovered. Scooped up from the mud also came objects of jade, gold, copper, ebony, along with balls of copal incense and obsidian knives, a treasure valued at two million dollars.

All of this pioneering underwater excavation was accomplished with maximum secrecy. When it was terminated "Don Eduardo" continued to live quietly at Chichén-Itzá pursuing his scholarly research over the years.

The second man was the governor of Yucatan in the early 1920s, Felipe Carrillo Puerto, a Mayan, a socialist and a driving, creative force in the region. He fought against oppression of the natives. He sought a restoration of the Mayan ruins, and of the ancient Mayan crafts, music, and dance as a means of instilling pride in his people.

Exciting archaeological finds had been made all over the world in this era, many by Americans, and all reported extensively in the American press. Still, little was known of the Mayan ruins in Yucatan. To change this Governor Puerto drafted plans to open up the area to exploration and tourism and invited foreign cooperation.

The Carnegie Institution was one of the first to respond and planned a preliminary survey of the ruins to be carried out in 1923 prior to major work at Chichén-Itzá to follow the next year.

The third person, drawn inexorably to the Sacred Well, was Alma Reed, a young San Francisco newspaper columnist with a personal interest in archaeology and art and a consuming passion for social justice.

In 1921, while on a visit to San Quentin prison, she saw on the bulletin board an invitation to the hanging of a sixteen-year-old Mexican boy who had killed the foreman of his work gang in a fight. Alma visited the youth and found he spoke no English and was confused over what was happening to him. Her front-page treatment brought a stay of execution and a change in the law upping the minimum age for capital punishment to eighteen.

The story was carried widely in the Mexican press. It resulted in an invitation to her from Mexico's President Obregón to visit the country as his semiofficial guest. She went, fell in love with Mexico, learned of its vast archaeological wealth, and met the leading Mexican muralists, both interests leading to later books.

While she was in Mexico the editor of the *New York Times* asked Alma to join the upcoming Yucatan survey group of the Carnegie Institution. *Collier's* magazine asked her to do a piece on Felipe Carrillo Puerto, the governor. It was another turn of the wheel.

When these two crusaders met they instantly felt a strong mutual attraction. During the two-week stay of the scientists, Felipe devoted much attention to Alma as he showed the visiting group the pueblos, his social experiments, the ruins, the dances, and the fiestas. At Chichén-Itzá Alma met Edward Thompson, now in his later years.

The old scholar asked Alma to return alone that evening, and he would give her an important story. She came back and Thompson took her to the edge of the Sacred Well, where they sat down. He quietly told her his "confession." He had recovered a half-ton of valuable Mayan artifacts from the Sacred Well. He had sent these, a few at a time, through the diplomatic pouch, to Harvard's Peabody Museum where they still remained hidden from public view. He could keep his secret no longer.

On top of this exciting information that same evening Felipe told Alma of his love for her. He asked her to stay in Yucatan with him, sharing in his fight to better the lives of the people as well as to write and study about the ancient Mayan past.

But Alma had to return to the United States. Before writing her "scoop," she was determined to go to Cambridge and verify the existence of the treasure. She took with her two Mexicans as witnesses from the New York consulate. After some confusion and delay at the Peabody Museum, the trio finally was shown the half-ton of artifacts. They were just as Thompson had described them, away from public view in an upstairs storage room.

Alma's story appeared in the *New York Times* magazine of April 4, 1923, as the most important discovery in the history of American archaeology. She was twenty-nine at the time. The story was a sensation. It sped around the world, launching her into further writing and lecturing about these Mayan ruins.

Adding to the furor was a lawsuit brought by the Mexican

government against the Peabody Museum. They asked for return of the artifacts or two million dollars. Thompson was dead by the time the Mexican Supreme Court settled in his favor. They ruled at the time of his excavation there were no adequate Mexican local or federal laws to prevent exportation of these artifacts which Thompson had found on his own property.

Capitalizing on this publicity, Felipe set out to hold what was billed as the greatest reunion of the Mayan race since the Conquest. There would be a ten-day festival in the 1000-foot-long ball court at Chichén-Itzá. The whole population of Yucatan was invited. Transportation and lodging would be provided. Central to the pageantry would be ancient Mayan dances, music, drama, and costumes.

The governor desperately wanted his love, Alma, there. She could not attend. The publisher of the *New York Times* had given her an important mission. He told her the paper was seriously giving consideration to supporting U.S. recognition for Mexico. But they needed proof of President Obregón's intentions for the future of Mexico. This would be revealed in his 1923 budget if she could obtain an advance copy. Alma was able to do this through her Mexican contacts. She wrote a story on the positive programs projected. Hearst with his chain was also thinking of backing recognition. Hearst asked the young journalist to do a series on constructive programs planned by the Mexican government after years of revolutionary turmoil.

With these stories behind her, Alma once again was able to head for Yucatan. The first American Continental Press Conference, at the governor's invitation, was to be held at Merida. At Veracruz she joined hundreds of journalists from all parts of the Western Hemisphere. They boarded a Mexican battleship bound for Progreso, Yucatan.

The trip was marred by a violent storm, a *norte*. In the midst of this, as the wind howled, Alma was serenaded outside her cabin by a quartet sent by Felipe. It was her song, "Peregrina," meaning "pilgrim," written at Felipe's suggestion by the Yucatecan poet, Luis Rosada Vega, with music supplied by well-known composer Ricardo Palmerin. Even today this remains one of the most popular songs in Mexico and is heard frequently throughout the country or wherever Mexican love songs are played.

At the first meeting the governor of Yucatan read a telegram to the journalists announcing recognition of Mexico by the United States. Alma was given credit for playing an important part in this.

Meanwhile, once again, the Mexican political pot was begin-ing to boil. President Obregón's term was coming to an end. He supported General Calles as his successor, also Felipe's choice. A rival presidential candidate was de la Huerta, who stirred up the planters and the church in Yucatan against the governor. Felipe tried to organize the military, but they had been bought off. He and Alma had become engaged. As the revolt spread, she was in San Francisco making final arrangements for their marriage.

Felipe, with his three brothers and six friends, sought to escape Yucatan in a small fishing boat. They were captured, given a mock trial, lined up against a wall of the Merida cemetery on January 3, 1924, and shot. As a final torture, the night before Felipe was killed, his jailers had mockingly sung "Peregrina" outside his cell.

For her reporting on the Sacred Well the Mexican government awarded Alma the Aztec Eagle, the highest award to be given a foreigner. She had become a well-known, respected, and tragic figure in Mexico where she had gone to live. We became friends in the late 1950s.

To some degree, because of her urging, at the fifty-eighth Congress of American Anthropologists held in 1959 the main artifacts Thompson had recovered from the Sacred Well were returned to Mexico. Resentment against Thompson still lived on in Yucatan. From my desk in the American Embassy I had joined Alma by recommending to our ambassador, the late Robert C. Hill, that he, on behalf of the U.S. government, back return of these artifacts. Today this treasure may be seen in the Museum of Anthropology in Mexico City.

Dr. Eusebio Davalos Hurtado, director of Mexico's National

Institute of Anthropology and History (INHA) authorized the 1961 CEDAM exploration of the Sacred Well of Chichén-Itzá. When interviewed on Alma's radio program, "What's New In Mexican Archaeology," he said, "Not even the surface of possibilities has yet been scratched."

This was the background for CEDAM's third expedition in 1961. And, of course, Alma would be there to report it.

6

Yucatan has no surface rivers. It is covered by a soft limestone cap. Rainwater seeps through this porous rock, and here and there, when a part of the cap collapses, it forms a sinkhole, or *cenote.* Such was the Sacred Well of Chichén-Itzá. One speculation was in the early days the Mayas had migrated to this region, drawn by the fresh water in these wells. They came to believe that Chac, God of Water, inhabited the depths of the Sacred Well.

Where did the Maya come from? There were many theories. Some said they had come from Asia, crossing the Bering Strait many centuries ago. Others said they had come from the Southwest Pacific. Some said they had originated in the Middle East. There was even a theory they had come from outer space.

Certainly they built one of the world's most sophisticated civilizations. They constructed magnificent buildings and cities and had advanced knowledge of such things as agriculture, mathematics, art, and astronomy.

Sometime, however, perhaps a half-century before Columbus discovered the New World, all this activity ceased to be when the Mayan cities were abandoned. Why they did this and where they went remains one of the world's major archaeological mysteries. Over the years the jungle took over. Restoration work on the ruins began in the 1920s, especially at Chichén-Itzá.

Pablo Bush, CEDAM's president, had long believed that in the depths of the Sacred Well might well be found major clues to the Mayan past, possibly even something like the Rosetta stone, found by the scientists who accompanied Napoleon on his invasion of Egypt.

In 1954, at Pablo's urging, members of Mexico's Frogmen's Club tried diving in the dark waters of the well. Visibility was nil and their time underwater was limited. They sought to ignore the legend of death coming to those who invaded these depths. This first effort was not a success. From this experience Pablo learned that divers alone were not enough to excavate the well. Special equipment would be needed.

His friend George Clark, of Chattanooga, Tennessee, gave him the idea. What was needed for this exploration was an airlift. This basically was a compressor on a floating platform. It had an eight-inch suction tube to vacuum the mud at the bottom. Divers guided the tube. Mud, stones, objects, whatever, all were drawn up the tube and deposited on the platform. Here the water was filtered through steel mesh, leaving behind the mud and its contents for examination.

The apparatus had been invented by Edwin Link, a CEDAM advisor, and had been used by him in salvage work at Port Royal, the city in Jamaica submerged in the sea in 1692 by an earthquake. Link agreed with the airlift concept and recommended Pablo to Norman Scott, an expert with the device. He was continuing the work at Port Royal while Link was investigating the Roman city of Caesarea under the waters of Israel.

A conference on the project was held in Mexico City with Dr. Eusebio Dávalos Hurtado, director of Mexico's Institute of Anthropology and History in the chair. Other INHA archaeological specialists were present. On the U.S. side were Melvin Payne, chairman of the board of the National Geographic Society, and Dr. Matthew Stirling, eminent North American scholar who had explored many archaeological sites in Mexico for the Smithsonian Institution. Pablo was there with Norman Scott, who had agreed to supervise the underwater operations.

It was agreed that a representative from INHA would be the top authority. Assisting Norman Scott would be five CEDAM divers, including two from the Mexican navy. Genaro Hurtado,

CEDAM's photographer, would be responsible for the underwater pictorial record.

Trucks, tents, the raft, airlift, and other equipment were taken to the Sacred Well in December 1960 and set in place. In January 1961 the work began. The top of the tree-lined circular rim of the well was around 180 feet in diameter. The white limestone walls dropped 85 feet to the surface of the pool. The bottom went from 30 feet deep, sloping off to a depth of 55 feet. Many large stones were on the bottom. It was thought these had probably fallen from a small ruined temple on the edge of the well. It was impossible to see in the black waters. Besides this, floating particles in the water reflected the light.

The divers worked in pairs. While one was on the bottom, the other stood on the raft holding a rope connecting the two. It was used as a signal to stop the compressor and the suction in the tube in emergencies. They were fearful they would dislodge large rocks that could roll down on the divers.

Guiding the mouth of the suction tube in the darkness was not without danger. The operator had to push the mud close to the mouth of the tube with his hands, being careful not to be grabbed by the powerful suction. In practice it wasn't the airlift or falling rocks that sidelined the divers—it was almost constant infections of the ears, nose, and throat, along with stomach upsets.

One of the first objects to be sucked up from the bottom was an underwater searchlight. It had been lost by Hector Mestre during the earlier exploration of the well by Mexican frogmen. Mestre was now CEDAM's treasurer. They took recovery of his light as a good omen.

Among the earliest artifacts recovered were two idols made of rubber, as well as a rubber ball. This was exciting. It had long been known that the Mayans played their ritual ball game with a rubber bouncing ball. Several of these had been sent by the early Spaniards to the King of Spain as a curiosity. The idols were unique.

Over the weeks thousands of artifacts were sucked up or

brought up by hand from the bottom by the divers: wood and rubber figures, jade pendants, coral beads, flint arrows, copal incense, obsidian knives, ceramics, gold ear plugs, copper rattles, grinding stones, and much more. Among other things they found bones of men, women, children, and babies, which seemed to refute the legend that only young virgins were sacrificed at the well.

After four months, with no advance warning, INHA suddenly ordered the work terminated. Authorities of this powerful guardian of Mexico's national treasures had decided the airlift procedure was damaging fragile artifacts. Besides, they concluded this technique made it impossible to assess the historical stratigraphy of the salvaged objects. Despite keen disappointment, there was no alternative but to comply.

Edward Thompson had written he believed his salvage represented less than 10 percent of the contents of the well. Pablo, looking down at the muddy waters, said he concurred. This estimate was still valid. The operation had been conducted at the upper levels. The artifacts recovered were from a period when the Mayan empire was in decline. Those huge stones on the bottom, Pablo said, could cover a treasure equaling that found in the tomb of Tutankhamen in Egypt.

Next time it would be absolutely necessary to remove those stones. One thing was certain—CEDAM did intend to come back.

7

In the summer of 1961, CEDAM set out to explore Chinchorro Reef. This large coral reef is situated about fifty miles out to sea off the southernmost coast of Quintana Roo near Mexico's border with British Honduras, now Belize. They also planned a general, rather relaxed, investigation south down the mainland coast from Cozumel Island.

For their exploration CEDAM assembled a "fleet" of seven boats, not very impressive any way you looked at it, but the largest ever in those remote waters. They had their old standby, the fishing schooner *Cozumel,* also the motor-sailboat *Pez Vela.* A little boat, the *Gallito,* acted as their floating cookhouse. American fisherman, guide, and writer, Emmett Gowen provided two outboard motorboats, *La Paila* and *Matancero,* while the Mexican navy sent the corvette *David Porter* and the patrol boat *A.M.5.*

Starting from Cozumel Island the group sailed down the coast, stopping briefly at sites of previous CEDAM adventures—Akumal, Matanceros, and the cove of Xelhá. At Tulum, the large Mayan walled city situated on the coast, the lone caretaker complained to Pablo of his inability to keep the weeds under control as he hacked away with his machete among the magnificent ruins. As an evidence of change, today Tulum is a carefully preserved archaeological zone visited by thousands of tourists.

Further south CEDAM came to the Boca Paila River, Gowen's special fishing ground. He called this his "domain." The area was full of fish—snappers, robalos, tarpons, bass, bones, jacks, and a variety of others. A few years earlier they had gone up the old Mayan canals to the lakes to look for the "Lost City." This time they went briefly to Lake Caapechén, where they examined on the shore a small, tomblike temple with a low entrance. In the opinion of Braulio Garcia, the archaeologist with the group, the door was built this way to force worshipers to enter the temple on their knees.

Much of CEDAM's information about shipwrecks and such came from legends, rumors, and old tales known to the natives. It was not easy for strangers to come by this information. Contacts had to be developed over time. Fortunately, Gowen was very *simpatico* and during his years of fishing in the area had made many friends.

One of these was Capt. Francisco Cantó, an old sea captain familiar with these waters. He had first given CEDAM information about the Matanceros wreck site. To see the captain the little fleet crossed Ascensión Bay to Punta Pájaros, so called because of the myriads of sea swallows to be found there. Now the old captain owned a coconut plantation planted among vestiges of Mayan ruins.

Pablo wanted to ask him something. It concerned the previous year when CEDAM was on Isla Mujeres. He had been discussing the old Mayan civilization one night with Father Bernardo Nagle, of the Maryknoll Order, who told him an intriguing rumor.

The good father said he had heard that at a place called Chunpom living "jungle Mayas" had a temple where they kept a mysterious "Book of God." Deep in the jungle of Quintana Roo, inland from Tulum, this sacred book was said to be their link with their ancient past, a legacy for succeeding generations. The priest said he had heard it was accessible only to initiated chiefs. The temple was guarded night and day by armed warriors.

At Pablo's request, while the CEDAM divers were working the wrecks, Father Nagle dispatched one of his Mayan converts from Isla Mujeres to see if he could get through the jungle to Chunpom. The message he carried to the head priest asked permission for CEDAM to photograph their temple and examine its contents.

Before CEDAM wound up its expedition the young envoy was back. He had a confusing and threatening message. He had been told if the outsiders were representatives of Queen Victoria of England they would be welcome. But if they were *huaches* (Mexicans) they had better stay away if they valued their lives!

Now, at Punta Pájaros, Pablo and the others asked Captain Cantó what he knew about this matter. The old captain did not disappoint them. Yes, he had heard of the temple and the "Book of God." This jungle community, he had heard, was under a chief or "Tatich," and there were some on Cozumel Island who claimed to know the family of this man. It was all very provocative, but this matter would have to wait.

It was time to explore Chinchorro Reef. They had a chart of the place drawn by an English cartographer in 1839. It had names, possibly indicating wrecks, like "Wreck Reef," "Skylark Ledges," and "Blackford Ledge." The reef was 70 kilometers long and 27 kilometers wide. Most of the reef was just under the turquoise water, jagged coral waiting to capture ships. Three widely separated islets were above water. These were Cayo Norte, Cayo Centro, and Cayo Lobo in the south, all said to once have been the lair of pirates.

First stop was Cayo Norte. They pitched their tents beside the ruins of a lighthouse destroyed the previous year by Hurricane Abby. Its replacement was an automatic one. The islet's only inhabitants were birds and crabs. The surrounding sea was transparent. On the bottom through their snorkels swimmers could see numerous giant conch, lobsters, sea worms, and the remains of five shipwrecks, none over eighty years old in their judgment.

They pushed on to Cayo Centro, 20 kilometers to the south. This was a 10-kilometer-long islet with a small lake in the center. It was a tangled mass of mangrove trees, the home of thousands of birds. Most unpromising. At the southern extremity of the reef they found the lighthouse there also had been destroyed by a hurricane and not replaced. This islet was so small as to be negligible.

Time was running out. As if on cue word came of the approach of a hurricane. They would have to move out. Not that it mattered much. This had been an almost lazy exploration. CEDAM had made new friends, maintained some old ones, poked around ruins, investigated the area, eaten a lot of lobster and fish, and dived repeatedly into the crystal clear waters.

But all was not lost. Something had come of it. Already plans for next year's expedition were being formed in Pablo's mind, and this one would have nothing to do with shipwrecks, diving, or the sea.

8

It was the American, Emmett Gowen, who brought matters to a head early in 1962. He wrote Pablo in Mexico City that while cruising his "domain" at Boca Paila he had met a young man who claimed to be the son of the chief, or "Tatich," of the jungle Mayas at Chunpom. His name was Martiniano Vega. He was a worker on a coconut plantation, and he had agreed to lead a CEDAM expedition to Chunpom and act as a liaison with his father.

Pablo quickly assembled a small group: Carlos Linder; Jim Webster, an American photographer; and Chuck Ennis, a San Antonio businessman, former hunting companion of Pablo's, now in love with conservation, exploration, and Mayan Mexico.

From Boca Paila they traveled by launch to Chunyaxche where tents were set up on the beach. A drover with a string of pack mules rented from a chicle station at Tancáh met them. Riding these beasts in the suffocating heat took the small band of adventurers two days to get through the dense jungle with its wild animals, poisonous snakes, and myriads of stinging insects.

They arrived half-sick at Chunpom. It was a collection of huts with thatched roofs. With the exception of Martiniano's father, who turned out to be Mexican, not Mayan, they found the reception definitely unfriendly. They had been misinformed. The name of the Tatich was Juan Bautista Vega all right. Although high ranking, something like secretary-general in the hierarchy, he was not the chief. The real chief, a sullen-eyed Mayan, kept the strangers apart from the people. He forbade them to go near or photograph the largest hut, which was the temple where they supposedly housed the "Book of God."

It was clear that the Tatich, who was seventy-six years of age, was seriously ill. A hernia kept him almost immobilized. He could hardly eat because of a tumor of the mouth. He was unable to walk or travel by mule. Without medical treatment he would soon die.

In several of the huts the visitors saw pictures of Queen Victoria on the walls as they passed. Later research told them in the early years the Spaniards had founded the city of Merida in the northern part of the Yucatan peninsula. To the south, early in the seventeenth century, the British had founded the town of Belize. Both towns became centers for the export of mahogany. Over the years there was fighting.

As time went by the Mayas sought to resist the encroachment of the Mexicans and reclaim the land for themselves. In the 1840s the War of the Castes broke out. Mayas were killing the Mexicans or *huaches* as they called them. They needed arms and these came, smuggled from Belize along with other supplies. Thus the reverence for Queen Victoria, whom they believed still to be on the throne. These Mayas living in the jungle had long memories. They still distrusted anyone from the outside world, especially the *huaches*.

Juan Bautista told them his amazing story and they taped it. It began in 1896 on the island of Cozumel where he was born. He was ten years old at the time when a missionary of unknown nationality came to the island. They called him Dr. Fábricas. He, too, had heard of the "Book of God" kept by the jungle Mayas and wanted to see it.

Dr. Fábricas needed a boat and a crew to take him to the mainland, but the people of Cozumel were afraid of the fierce Mayas living in the jungles of Quintana Roo. In the end Juan's stepfather, Ruperto Loria, agreed to make the trip, taking along Juan and a sailor named Ignacio Medina.

The small boat was loaded with supplies, including some gold with which the missionary hoped to buy Mayan friendship. When all was ready they shoved off, sailed over to the coast, and landed in the early afternoon on the beach by the walled city of Tulum. All was quiet. Ruperto busied himself cleaning some fish he had caught at sea on the way down. Though he had no gold, he thought to offer these as a gift. No one appeared.

Early next morning, with the beach still deserted, Ruperto made preparations to sail back to Cozumel with young Juan and Ignacio. Dr. Fábricas would be left behind as they had agreed. Suddenly three armed Mayas appeared. Silently they crossed the beach, ransacked the boat, and began stuffing themselves, eating everything they could find.

When they had finished eating, still not saying a word, they grabbed the sailor, Ignacio, and beat him to the ground with their guns. Then, before their horrified eyes, one of the Mayas cut Ignacio's throat with a machete. They tied up the others.

More Mayas emerged from the ruins. Chattering among themselves, they divided everything on the boat before filling the small craft with coconut husks and driftwood and setting it on fire. Herding the prisoners before them they set off for Tulum.

Before they had gone very far, without warning, several began beating Dr. Fábricas, and the unfortunate missionary was finished off in the same manner as Ignacio. Next it was Juan's stepfather's turn. They forced him down on the beach and cut off his head with a machete. All the Mayas were nearly naked. They stripped the clothes from Ruperto's body and divided the pieces.

One of the Mayas raised his machete as though to kill the youngster, but his hand was stayed, Juan said, "by the hand of God."

They took Juan through the jungle to Chunpom where outside one of the huts he was tied, fed only refuse, and kept literally like a pig. Unlike a pig, the youngster was subject to the jeers and blows of the Mayan boys. He would have died, Juan said, but what saved him was Dr. Fábricas's Bible.

Somehow, amidst all the bloodshed on the beach, he had been able to pick up and hang on to this holy book. One of the temple priests noticed Juan reading his Bible while tied in the pigsty. If the boy could read Spanish this could be useful.

He took him to the chief, where it was decided he would be given guarded liberties in exchange for teaching Spanish to twenty-five Mayan warriors in training to battle the Mexicans.

They had kept him as a prisoner for sixty-six years. During this time, he had married a Mayan woman and they had had three children. He had risen to a peculiar position of power. They had made him Tatich, secretary-general or counselor. As an outsider they had used him as a peacemaker among the contending Mayan factions.

Fighting continued between the Mexican government and the jungle Mayas. It was a vicious war with cruelty on both sides. The government brought in dogs trained to discover the Mayan traps laid in the jungle for the Mexicans. There was much bloodshed.

When a pact between the opposing forces was finally negotiated, Juan, the Tatich, the Mexican prisoner of the jungle Mayas, became involved as a middleman. In return for his services he was named a general in the Mexican army by the Mexican president. The Mexican government wanted to take Juan to Mexico City to receive his uniform and commission, but the Mayas would not let him go. They feared treachery on the part of the *huaches*. Besides, they didn't want to lose their Tatich, who had become so valuable to them in negotiating differences not only with the Mexicans, but also among themselves.

But, Juan told the CEDAM explorers, all that was long ago. Now he was old, seriously ill, and still a prisoner. The night before they arrived he had had a dream help would come.

9

The decision was made at Chuck Ennis's suggestion. They were going to save the old Tatich. This CEDAM venture was dubbed "Operation Rescue." Pablo, back in Mexico City, contacted his friends, Gen. Roberto Fierro, chief of the Mexican Air Force, and Gen. Augustin Olachea, secretary of national defense. Years earlier the latter had fought against the jungle Mayas in Quintana Roo as Mexico sought to pacify and integrate this region into the Republic. They agreed to send a plane and a helicopter.

While waiting for these Pablo dispatched a small plane from the Aero Safari Service. He had formed this with Capt. Robert Fierro, nephew of the head of the air force. It was used for jungle hunting trips, to connect Cozumel with Tulum. The plane was sent to fly over to the mainland and locate Chunpom from the air as well as to locate a place where the helicopter could land. The following day Captain Fierro led the military helicopter to the site. Several of the expeditionaries were aboard. They had with them Argimiro Argüelles, a relative of the Tatich from Cozumel. He was to act as an intermediary.

The helicopter landed close to the village. The jungle Mayan inhabitants, though curious, were kept back by their chief. Only the old Tatich alone made his way forward. He was barefoot and dressed in the local attire of white shirt and trousers. The old man, sick and ailing, painfully made his way to the chopper where, on learning that Argimiro was his brother-in-law, he embraced him.

Juan was asked if they might visit the temple. He agreed

to go with them but asked that they refrain from making any photographs at least until he could judge reaction. At the entrance to this palm-thatched building, they were stopped by a guard armed with a machete. Juan Bautista had told them the temple was indeed guarded day and night by a force of about twenty men who changed their watch at regular intervals. Some had old rifles; most had machetes.

The guard stood with his hand on the hilt of his machete and said he would not let them pass. Juan told the fierce young man these were his friends. The guard replied it was his sacred duty to kill any *huache* who tried to enter the temple. Then he seemed to relent. They could pass if Juan would give the order. Bautista replied he would not do this. The guard was right. The temple was under his care. They did not see the inside of the temple or the "Book of God."

A little distance from the temple they sat down to talk. Juan would tell them little about the "Book of God." Yes, he had read them. There were three books, not one. They contained much information about the history of the Mayan people. The books were holy. They had miraculously appeared one evening long ago surrounded by candles in a hut at the entrance to the town. The contents were in an antique Mayan language. The books were of a paper made from bark. That was it. He would say no more.

The old man was told that CEDAM wanted to take him to Mexico City for medical treatment. Juan agreed provided he could take along his son, Martiniano. Also, he would have to talk to the Council of Chieftains. After all, he was still a prisoner. The group waited. Presently he returned. The chieftains had agreed, he said, only after he had persuaded them that in Mexico City he would be able to plead with the Mexican authorities to help them. They had given their permission reluctantly.

Juan asked the CEDAM group to return in one month. He needed the time to contact the chiefs of the other scattered tribes and get their permission. He was told "Now or never." The expedition was costly. It would be impossible to delay. Juan then asked that they return the next day. Argimiro was left to help contact the other jungle chiefs. The next day, when the helicopter returned, Juan was ready to depart.

From Chunpom the group returned to Cozumel. The old man went in a Mexican air force plane to Mexico City where Pablo met him with a small CEDAM group. They were met by the chief of the Mexican Air Force and senior members of his staff.

Arrangements had been made for Juan Bautista and his son to stay in the military hospital. They were lodged in one of the general's suites. This had been arranged by the secretary of national defense in recognition of the past rank given Juan.

The Tatich was overwhelmed and grateful. He felt he had been snatched from the jaws of death. Before anything else he told Pablo first he wanted to go to the Basilica of Guadalupe to visit the Shrine of the Virgin, Mexico's patron saint. Pablo and a select CEDAM group took him there.

The priest in charge of the shrine took him to the main altar where he stood transfixed, looking at the Virgin, talking to her in Mayan. Tears came to the eyes of the CEDAM group. Then he said: "I am ready. Do to me whatever must be done."

The hernia operation and treatment for the tumor on his mouth were successful. They also repaired his teeth, performed an abdominal operation on his son, Martiniano, and operated on the younger man for cataracts.

Altogether Juan remained in the hospital for three months. During this time there was a steady stream of visitors. High-ranking political and military figures came to visit the old Tatich. A member of the British Embassy came by with a photograph of HRM Queen Elizabeth as a gentle hint that Victoria was no more.

One visitor, Esteban Bakle, was an old veteran who had read about Juan Bautista in the press. He had come all the way from Jalisco to see him. Years earlier this man had been captured by the jungle Mayas during the fighting. He said Juan Bautista had saved his life. He wanted to thank him.

Even the secretary of defense, Gen. Augustin Olachea, dropped by to exchange stories of the old battles.

When the time came for his release, Pablo took "Gen." Juan Bautista Vega to call on the governor of Quintana Roo at his residence in Mexico City. The governor told them he had been instructed by the president of Mexico to act favorably on requests for assistance to the jungle Mayas conveyed by their emissary, Juan Bautista.

His requests were simple. He said they wanted the road at Felipe Carrillo Puerto, old capital of the Mayas before the Mexican takeover, extended to Chunpom. Besides this they wanted school materials and musical instruments for their band, as well as something done to insure a better supply of fresh water.

As for himself, Juan told Pablo, there was one thing he wanted and that was a chance to visit with his relatives on Cozumel Island from whence he had departed at the age of ten. This was arranged and he stayed there for a year.

When this time was up there was no helicopter available to fly back into Chunpom. CEDAM had brought Juan out and they would take him back. Pablo formed a delegation with Genaro Hurtado, Chuck Ennis, Roy Pinney (a writer), and Col. Richard Harris, air attaché of the American Embassy. Once again they had to use a mule train, wending their way slowly through the steaming jungle.

Again the reception was unfriendly, not only for the strangers, but even the Tatich was received coldly by the chieftains. This despite the musical instruments, school supplies, and the promise of government help with the road and fresh water.

Juan told the CEDAM group sadly the jungle Mayas thought he had sold out to the *huaches* and would not be back. During his absence they had disposed of his few belongings. Nevertheless, he said he would remain in the jungle with them. He had no doubt he could once again win the support of the chieftains. It would just take time.

The old man lived until 1969. He died at almost the same time as the last Mayan general, Francisco Mai. They were buried side by side in the churchyard at Carrillo Puerto. CEDAM and the Maryknoll fathers erected a memorial to these two men.

In 1979 Pablo again visited Chunpom, now a twenty-minute drive south of Tulum on a paved road and then eight kilometers on a good surface dirt road. He found the village basically unchanged except for the road, a new school, and a water tower linked to a well giving fresh water.

The same head priest was there but now less hostile. Pablo and his wife, Elsy, were permitted to enter the temple and pray. Before entering they were asked to take off their shoes. Inside there was a large room for the people to hear mass and pray. There was also a small room, an inner sanctum. In this they found an altar with a table, a large chair, and a small chair. These were for the Father and the Son when during festivities food was placed on the table.

Many candles burned before the figures of saints on the altar. They learned that many of these images had been stolen during the War of the Castes from destroyed churches. These images were carefully guarded, for the jungle Mayas believed if ever they should lose these saints terrible calamities would befall them.

Pablo and his wife left still, after all these years, without seeing the "Book of God."

Before leaving, the Mayas asked Pablo for a coronet and a bass drum to complete their band. He promised he'd get them the instruments.

10

Due in large part to Pablo Bush's skillful promotional efforts, CEDAM—its concepts and accomplishments—was reaching out, not so much to the general public as to those persons, and especially their leaders, interested in underwater exploration when this activity was still very new.

In the fall of 1962, Pablo went to London to attend the Second World Congress of Underwater Activities. This meeting had been organized by the British Sub-Aqua Club in association with the Confederatione Mondiale des Activites Subaquatiques (CMAS) founded by Commandant Jacques Yves Cousteau.

This was very top drawer. The official host was HRH Prince Philip, Duke of Edinburgh, himself an ardent diver and president of the British Sub-Aqua Club.

CEDAM was to be sponsored for membership in the Cousteau organization. Sponsors were Walter Feinberg, president of the Boston Sea Rovers, one of the most distinguished diving groups in the U.S., and by Dave Stith, of the Philadelphia Depth Chargers. Stith recently had been elected president of the Underwater Society of America.

At the congress Pablo met and became friends with Dr. Andreas B. Rechnitzer, an outstanding American oceanographer who later would become president of CEDAM International.

Much of the world's attention was increasingly being drawn to outer space. Commander Cousteau described the areas beneath the sea as "inner space." He said they demanded urgent attention.

Talks and discussions focused on the state of the art of underwater exploration and plans for the future. Topics covered such objects as archaeology, marine biology, zoology, geology, underwater photography, sports, equipment, training, and the medical and legal aspects of diving.

There was a danger of CEDAM being overlooked in such outstanding company. But luck was with Pablo. On the second night of the congress, completely by happenstance, ABC-TV's film, *The Sacred Well of Chichén-Itzá,* was shown on British television. This film was based on the excavation in which CEDAM of Mexico had played a leading role. And, of course, many of the delegates had read an earlier article on this exploration in *National Geographic.*

The stage was set. Pablo had brought his own copy of this film with him. As part of his presentation he screened seventeen for the delegates. In his talk he gave them greetings from Dr. Eusebio Davalos Hurtado, director of Mexico's National Institute of Anthropology and History. He announced that Mexico would welcome work-study diving teams of students and teachers to investigate archaeological zones under the guidance of Mexican specialists. CEDAM of Mexico, the newest member, began receiving much attention.

A number of European delegates wanted to borrow the CEDAM film. Plans were made to circulate it to Italy, Greece, Germany, Finland, Switzerland, and other countries. Pablo later showed a Spanish-language version to the 500 members of Spain's Asociacion de Pesca Submarina in Barcelona. Afterward this print was loaned to the Argentine delegate to tour the diving clubs of South America.

Jacques Dumas, president of CMAS, told Pablo that his organization already had thirty-three member nations. He commissioned CEDAM to serve as the organizing body for CMAS in Mexico, Central America, and the Caribbean.

When he returned to Mexico, Pablo in his report to the CEDAM membership said: "The importance of this conference cannot be overestimated. It points out clearly and on a worldwide level that diving is not just a sport, but a new and potent arm in the scientific community. In man's constant struggle to know his origins and to plot his future, the diver now is acclaimed as a vital and competent servant. From CEDAM's standpoint, the World Underwater Congress served as a forceful stamp of approval to the independent program of service and investigation that we have been promoting, in our own small way. For CEDAM, it also serves as a challenge for growth."

One possibility for growth came, of all places, from CEDAM's contact with the old Tatich. Pablo and CEDAM members had been cooperating with Station KTRK-TV, an ABC affiliate in Houston, on a reenactment of the story of Juan Bautista Vega and his life with the jungle Mayas. On conclusion of the filming, Pablo was invited to Houston. Among other honors he received a key to the city from Houston's mayor, Lewis Coutrer, and a portrait of Juan Bautista to be placed in the CEDAM Museum.

Intrigued with CEDAM's work, a number of prominent Houstonians asked how they could participate more directly. Pablo, Alma Reed, and Chuck Ennis talked about it. From these discussions came the idea for CEDAM International. The initials, they thought, could stand for Council for Exploration and Depth Archaeology in Meso-America.

In this preliminary thinking they thought this would be an organization in the educational field, sponsoring on a broader and larger scale work and programs initiated by CEDAM of Mexico. There were no commitments, but Pablo told the interested Houstonians formation of CEDAM International was a program that merited continuing study and discussion.

Meanwhile, CEDAM continued to branch out in Mexico. In 1963 members of the organization were the first to dive in the Sacred Lake of the Volcano of Toluca (Xinantecatl), establishing a record for diving at high altitudes (13,828 feet). Working with Mexican navy specialists and Dr. Miguel Guzman Peredo, head of GAISA, they developed special decompression tables for diving in elevated lakes such as this.

Pre-Colombian pieces of copal incense in the shape of the volcano, potsherds, a complete pottery vase, and a piece of wood carved to represent a streak of lightning were recovered.

Other items brought up caused speculation that some had been carried to Mexico from Spain during the Franco revolution on the yacht *Vita,* rumored to have carried away part of the country's gold. Found were a gold ring, gold false teeth, tin boxes inscribed with *Monte de Piedad de Madrid,* watch movements, a music box, and a church reliquary.

CEDAM members planned to return one day. What was needed in exploring the bottom of this high altitude lake was an airlift such as the one they had used earlier at Chichén-Itzá.

Some of the divers speculated that this lake, located not far from Mexico City, could possibly be sacred to the Aztec nation. Who knew what treasures might repose in the mud on the bottom?

11

In 1964 Pablo attended the Fourth Annual Convention of the Underwater Society of America in Philadelphia, where he pulled off the coup of persuading the society to have its convention the following year in Mexico City with CEDAM as the sponsor. Not only did this bring an enormous amount of prestige to CEDAM, but the society also benefited. With Pablo acting as convention chairman, assisted by Walter Feinberg of the Boston Sea Rovers, this turned out to be the most successful meeting the society had ever had.

The inaugural address by Mexico's secretary of the navy, Adm. Manuel Zermeno Araico, was made in the name of Pres. Adolfo Lopez Mateos. The event drew many of America's biggest names associated with underwater exploration. A review of the speakers and their subjects gives a good indication of progress being made in the new world of "inner space" by the mid-1960s. For this reason the program is listed in its entirety.

The convention opened with a symposium on "Manned Undersea Stations," chaired by James Dugan, president, Academy of Underwater Arts and Sciences. Participants were: Edwin A. Link, Sea Diver Corporation, "Equipment and Procedures for Sustained, Deep Dives to 400 Feet." Capt. George F. Bond (MC) USN, director, Submarine Medical Research Laboratory, U.S. Naval Submarine Base, New London, Conn., "Project Sealab I, Scheduled to be Occupied This Year." Commandant Jean Alinat, associate director, Musee Oceanographique, Monaco, "Cousteau's Continental Shelf Station Program."

The next day a morning session, "Underwater Archaeology," was chaired by John Huston, president, Council of Underwater Archaeology. Speakers were: Dr. Eusebio Davalos Hurtado, director, National Institute of Anthropology and History, "The Sacred Well of Chichén-Itzá and Its Contribution to Underwater Archaeology." Prof. Demetrio Sodi Morales, "Mayan Mythology and Their Aquatic Gods." Rear Admiral Armando Cañizares (to become CEDAM's second president), director of social security for the Mexican navy, "Pre-Spanish Navigation and Commerce in the Caribbean and Its Contribution to Underwater Archaeology." Jesus Bracamontes Avina, naval archaeologist, president of the Mexican Association of Navy Modelism and Maritime Culture, "Early Ship Construction in New Spain." CEDAM's historian, Alma Reed, "Aspects of Submarine Archaeology." At CEDAM's initiative a panel discussion was held on laws and ethics relating to wreck sites.

The afternoon session heard Mendel Peterson, head curator, Dept. Armed Forces History, Smithsonian Institution, "Spanish Plate Fleets." Lic. Jorge Gurria Lacroix, secretary, National Institute of Anthropology and History, "Trade Between Spain and New Spain." Marian H. Sagan, research assistant, Council of Underwater Archaeology, "Shipworms and Ship Preservation." Robert C. Wheeler, Minneapolis Historical Society, "Relics from the Rapids: Archaeology of the Fur Trade." Lon W. Mericle, research associate, anthropology, Milwaukee Public Museum, "Treasures of the Guatemalan Sacred Lakes." Donald P. Silva, assistant professor, Institute of Marine Science, Miami, "Dangerous Fish." Genaro Hurtado, CEDAM photographer, "Mineralogy of the Sacred Well of Chichén-Itzá."

On the same day, while the above was going on, an alternate symposium was held, chaired by Walter Feinberg, executive secretary, Academy of Underwater Arts and Sciences. Speakers were: John C. Bender, inventor and amateur archaeologist, "Underwater Metal Locator." Robert F. Dill, U.S. Navy Electronics Lab., San Diego, "Rivers of Sand, Cabo San Lucas, Baja California." Michael C. O'Neill, district manager, Dacor

Corp., "The Advancement of Scuba in Commercial Diving." Csaba Sziklai, U.S. Public Health Service, "Underwater Tests of Spatial Orientation." Richard Vahan, *Boston Herald Traveler,* "Diving Ethics: A Responsibility." William A. Newman, geology and meteorology instructor, dept. of natural science, Northeastern University, Boston, "Geological Investigation as an Aid to Underwater Archaeology." Ricardo H. Presbitero, chief of the dept. of underwater engineering, University of Mexico, "Sub-Aquatic Engineering." Dr. Aniceto Ortega, head of the Center of Sports Medicine, University of Mexico, "Sub-Aquatic Engineering in its Medical Aspects." Ricardo O. Bastida, biologist, "Application of Diving to Biological Studies in Argentine Waters." Dr. Miguel Guzman Peredo and Dr. Gaston Ezquerro Madrigal, Grupo Alpino y de Investigaciones Subacquaticas, "Diving at High Altitudes." Armando Yanez Correa, geological engineer, "A Geological Study of the Caribbean Coasts." B. M. Fisher, British Sub-Aqua Club, "Activities of the British Sub-Aqua Club, Present and Future." Ron Church, photographer, "Stressing Modern Equipment and How to Sell Your Pictures."

On the last day of the conference, Dr. Andreas Rechnitzer, program manager, Oceanology Programs, Navigation Systems Division, Autonetics, North American Aviation, chaired a symposium on "Organized Diving" with five topics: diving for science, training, sport, conservation, and diving legislation.

Speakers were: James Stewart, Scripps Institution of Oceanography, "Cooperative Scientific Diving Programs." Charles Turner, dept. of fish and game, San Pedro, California, "Special Population Surveys Using Lay Divers." William A. Newman, oceanographer, University of California, San Diego, "Interdisciplinary Use of Oceanography Relating to Submarine Archaeological Sites." William L. High, Bureau of Commercial Fisheries, Seattle, "Role of Organized Diving in Fisheries Research." Albert A. Tillman, executive director, International Underwater Film Festivals, "Organized Instructions." Ralph N. Davis, president, International Underwater Spearfishing Association, "Competitive International Underwater Spearfishing." Donald P. Silva, "Diving

Legislation." Bill Barada, *Skin Diver* magazine, "Maintaining the Right to Dive."

In an alternate symposium chaired by Walter Feinberg speakers were: Dr. Eugenie Clark, director, Cape Haze Marine Laboratory, Sarasota, "Recent Studies on Sharks, Particularly the Tiger Shark." Luis Marden, *National Geographic,* "Underwater Photography." Roy Pinney, photographer, "Caves and Their Relation to Underwater Archaeology." Harold E. Edgerton, professor, Massachusetts Institute of Technology, "Photography and Sonar for Underwater Projects." Nixon Griffis, U.S. Representative, Confederation Mondiale des Activites Subaquatiques, "Information Sources in Oceanography." Rear Admiral E. A. Stephan, USN, former oceanographer, USN, Chairman Deep Submergence Systems Review Group, "Work Done by [His] Committee."

Films were shown every night, including, of course, those CEDAM had made. At a final banquet CEDAM presented special awards. All in all it had been a great event for CEDAM. Now experts in every sort of underwater activity were familiar with CEDAM—its aims and accomplishments.

But Pablo still had one more card to play. For those who wanted to go, following the conference, he had planned "An Adventure in the Mexican Caribbean." A total of fifty went, representing the U.S., Argentina, Guatemala, Canada, the Bahamas, Italy, and England. Two transport planes from PEMEX, one from the Mexican air force, and two private planes, flew the group to Cozumel Island, still little known by tourists.

Andy Rechnitzer, one of those to go, said, after three days of staying up all hours at the conference when they arrived at sundown in Cozumel, he was more than ready for sleep. But Pablo couldn't wait. He told Andy to grab his toothbrush and bring just the clothes on his back. He and some of the others were off in a boat by moonlight.

Andy said later: "I found Cozumel was a startling place. You could look down to the bottom of twenty-five feet of water in the moonlight and see the sea urchins. And the water temperature was such you couldn't really tell when your hand went out of the air and into the water." That's when he first began to think seriously of working with CEDAM.

Quintana Roo was still desolate and without coastal roads. Pablo led the group of experts by boat to Akumal, to Xelhá, to Tulum, and to the old wreck of Matanceros where they dove and found some "goodies."

CEDAM's chief diver, Alfonso Arnold, went down off Cozumel and at depths ranging from 120 to 200 feet found the first black coral discovered in the area, a substance much prized by jewelers.

When the underwater experts had departed, Pablo knew he had accomplished his objective. The word about the Mexican Caribbean and its lure for divers would be spreading.

12

During World War II, the United States built a fine runway on Cozumel Island as part of the regional defense system. Afterward this was deserted except for an occasional plane landing. When CEDAM first came to the island in 1959, the population of the town of San Miguel was around 2,500 people. Thick jungle covered most of the island. There was only one hotel, Cabañas del Caribe, owned by Raul Gonzales, where CEDAM made its headquarters.

Following CEDAM's 1959 expedition to the Matanceros and the widespread publicity resulting from this exploration, word began to get around that Cozumel Island was the latest "in" place for divers. Tourism began to build. The second great takeoff for tourism was to come as a result of further publicity from Pablo's book, *Under the Waters of Mexico* and the Underwater Society meeting in Mexico City.

Cozumel Island, twenty-eight miles long, ten miles wide, meant "Land of the Swallows" in Mayan, and these early visitors built various temples and structures there. The first foreign visit was in 1518 by Juan de Grijalva. The second was made by Hernando Cortes the following year while on his way from Cuba to the conquest of Mexico.

For five years Cozumel Island had served as the rear base for CEDAM expeditions along the mainland coast and to the offshore reefs. Now, in the summer of 1965, Pablo thought perhaps Cozumel Island itself deserved some investigation. An expedition was planned to do this, specifically to explore several *cenote*s back in the jungle, one of which they had sighted from the air. Perhaps these, too, were held sacred by the Maya.

Arrangements were made for chicleros to use their machetes to hack a trail through the jungle to these *cenote*s. CEDAM would operate from a base camp set up at El Cedral, a village in the interior of the island.

But first there was other work to be done. Once again the fishing schooner *Cozumel* was charted. Accompanied by a Mexican navy cutter, the group sailed over to the mainland to camp under the coconut trees on the beach at Akumal. Tents were pitched, jungle hammocks slung, and spear fishermen went into the water to look for supper.

The youngest member of the group, Michael Blair, a teenager, later wrote how this underwater world had looked to him as a snorkeling spear fisherman when he saw it for the first time:

On the bay inside the reef "the water was as flat as a duckpond with the waves breaking on the white sand with little more than a gurgle." He entered the water, left the sandy bottom behind, and approached the coral gardens. "Large trees of red and purple fan coral swayed rhythmically to the surge of the waves. The brain coral, white and textural, grew to immense sizes, and the rough stick coral thrust itself into forms more diabolic than the work of the most imaginative of modern sculptors. And among it all the multicolored fish darted, slid, and fluttered through the clear water as their various metabolisms directed."

He said it seemed strange a world so different from our own existed just below the surface of the water. "One wanted to linger a day, a month, but we had to head for deeper water for the large fish."

Beyond the reef he found "a terrain of canyons and caves where all colors had been washed down to a green-blue, but where the visibility is, if anything, better than in the shallows." They returned to camp with a dingy full of fish, lobster, and one turtle.

The group's first task was at the Matanceros wreck. This was to raise a nine-foot, coral-encrusted cannon. Pablo had made previous arrangements to have a special tank built to receive this specimen. It was not easy. The effort to raise the cannon almost capsized their boat. Afterward the cannon was destined to spend three years in various chemicals while being treated in the tank so as not to decompose in the air.

Before returning to Cozumel Island the CEDAM divers briefly visited Xelhá inlet where they sighted a manatee, the seagoing mammal said to be the origin of the mermaid legend because under certain conditions in murky water the shape was vaguely like a woman's. Unfortunately, the local fishermen considered these rare and endangered creatures to be excellent eating.

The rumor of a wreck drew them south to Tulum. An immense waterspout suddenly appeared off their starboard bow. The helmsman made a hard turn, and the swirling mass of water passed with less than twenty yards to spare and they shot movies to prove it. This convinced them it was time to return to Cozumel.

There the jungle trails had been cut and the mules were waiting. It was known that for 400 years the Mayas had made annual pilgrimages to Cozumel to the shrine of the moon goddess Ix-Chel. There were numerous archaeological sites. After discussion and research, Pablo had selected three *cenote*s for investigation—Chen-Pita, Chen-Cedro, and Chen-Pon—all on the southern end of the island.

They turned their attention first to Chen-Pita. As they moved in the dim light of the jungle where the sun was shut out by the dense vegetation, they saw many butterflies and brightly colored birds. Underfoot large iguanas fled at their approach.

Young Blair described this *cenote:* "The large open expanse of water is continued under a cathedrallike overhang with a great stone bridge spanning it at one point. Cramped tunnels lead into further aquatic chambers that we dared not enter because our only source of light was a small and very cheap torch that I had to carry between my teeth when swimming. We decided that beautiful though it is, Chen-Pita was not sacred, so we determined to devote all our energy to the other two *cenote*s."

At first they tried reaching the other *cenote*s with the mules, but this had to be abandoned when it was found the sharp stones and numerous holes underfoot made the going too tough for even these rugged animals. Veteran diver Alfonso Arnold decided the only way in was for the divers to carry their own equipment.

Staggering under the weight of the tanks and the other gear took two hours of walking each way. This limited diving time to three days. After a brief investigation they also gave up on Chen-Pon. Visibility was less than a foot and the bottom was covered by ten feet of soft mud.

The focus was now on Chen-Cedro a mile further on, an unimpressive ten square yards of clear water under an overhang of rock. To reach the water the divers had to build a rude ladder. From this, when they toppled into the water, the vampire bats would swarm out from the overhang twittering in alarm.

Below water was a steep incline. They erected a lifeline and clung to this, shining submarine torches as they made their descent. At the bottom they found a jumble of enormous stone slabs. These had evidently been toppled into the water by natives searching what was probably a small Mayan temple on the side of the *cenote*. Any movement by the divers on the bottom stirred up mud. They were afraid too much digging beneath these slabs could start a landslide.

Nonetheless, the CEDAM divers recovered thirty items of stone and jade, a nose pin, a tool, ornaments, knives, and a phallic symbol. They also found nuts that possibly had been used for necklaces and a fragment of a human jaw. On the basis of this evidence, they concluded this was a well of sacrifice.

Perhaps of more importance than the discovery of these artifacts was their determination that this *cenote* contained fresh pure water. Pablo reported this finding to the authorities, grateful for any new source of potable water in this rapidly developing tourist resort.

So far 1965 had been a good year for CEDAM. Before it was over, there would be both a positive and a negative happening. On the positive side, giving the organization further publicity, was the final screening in Houston of the TV film made with

CEDAM cooperation on the story of Juan Bautista Vega. A TV newspaper critic said the film had caught "the essence of a strange and mystical story" and that it was "one of the most beautifully photographed, imaginative specials in local TV history."

The negative happening came in September in Mexico City. A school was to be built on a vacant lot next to the CEDAM museum Pablo had established. Workmen had excavated right up to the property line, weakening the foundation of the museum. Heavy rains came and the museum roof suddenly collapsed. In a minute objects Pablo had been collecting for twenty-five years, including ceramics worth probably $150,000 plus many of the CEDAM artifacts from under the sea, were destroyed, buried in the debris.

Andy Rechnitzer, who had been on the Cozumel *cenote* dives a few months earlier, wrote expressing the sentiments of many: "Those of us who were privileged to see the collection," he said, "feel personal distress."

Pablo said never mind—they would start again.

13

Oceanographer Andy Rechnitzer had become hooked on CEDAM. Like others he wanted the organization to have a better link with the United States and the international diving community. The idea had been kicked around for sometime. In 1966 something was finally done about it with Alma Reed and Chuck Ennis taking the lead. CEDAM International was born as a nonprofit organization. The acronym stood for Conservation, Exploration, Diving, Archaeology, and Museums. The word "Exploration" was later changed to "Education." Pablo Bush was the first president, succeeded the following year by Andy Rechnitzer.

CEDAM International was lucky to have Andy at the helm. As a distinguished scientist, he had been long involved in a broad range of ocean exploration all over the world. He held the world's depth record of diving in 1959 to 18,150 feet in the bathyscaphe *Trieste,* a record subsequently replaced in the same submersible by Lt. Don Walsh and Jacques Piccard, diving in the Marianas Trench, to the maximum depth possible of 35,800 feet.

At the White House in 1960 Andy received the highest navy department award, the Distinguished Civilian Service Award, from Pres. Dwight Eisenhower, both for his record dive as well as his leadership as scientist-in-charge of the NEKTON project that had developed the *Trieste,* now on exhibit in Washington, D.C., at the Navy Museum of History.

On assuming office Andy said: "To perpetuate the ideals

of our founder and continue extending the talents of our members to the service of science and mankind, CEDAM International will devote most of its program efforts to underwater exploration leading to the scientific recovery, conservation, and study of items that will provide a greater knowledge of mankind through the legacy available to divers in the underwater repositories of the world. We are now, by charter, an internationally oriented activity, and we do intend to expand the efforts of CEDAM International to all parts of the world."

In the summer of 1966, the year before all this, Andy was in Merida when Pablo told him the CEDAM of Mexico group was coming in that afternoon and that they should go to the airport to welcome them. Greetings were enthusiastic. Chief diver Arnold and the others urged Andy to come along; they were headed for Alacran Reef. Andy said he was not prepared, he had no equipment, but they said come along, we'll share with you.

The result was Andy found himself that night on the boat they had hired sleeping with his head on a coal sack. Later he said it was one of the best expeditions ever. A number of wreck sites were located. Consequently, in the summer of 1967, he organized "back-to-back" expeditions with CEDAM of Mexico and CEDAM International teaming up to continue the Alacran investigations.

Alacran, or "Scorpion," Reef was about 65 miles north of the Yucatan peninsula out of the small port of Progreso, semicircular, a mass of coral heads about 14 miles long and 8 miles wide, a death trap for ships over the centuries.

First target was the remains of a vessel discovered the previous year by Alfonso Arnold, co-leader with Rechnitzer of the 1967 expedition. The finds were exciting—four anchors and two cannon. Also at a depth of ten feet the divers found six cobblestones used for ballast.

Through later research they estimated these artifacts had come from a two-masted caravel about 150 feet long, carrying four cannon, four anchors, and a crew of about 200. The cannon were a breech-loading swivel type known as Culebrina Lombardos or "murderers." Each was 11½ feet long, weighing 300 pounds.

One couldn't be sure, but it was believed the ship might have been part of Adelantado Francisco de Montejo's armada from Spain, conqueror of the Yucatan in 1549 following a bloody campaign against the Mayas that lasted twenty-two years.

An interesting piece of information turned up in research was that two crew members of this armada had escaped death earlier when shipwrecked on reefs at Jamaica, called Vivoras, and had drifted to the Quintana Roo coast. One of these, Geronimo de Aguilar, joined Cortes in fighting the Mexican Indians. The other, Gonzalo Guerrero, wound up living with the Mayas in Chetumal, Quintana Roo. He married a Mayan woman and they had children. These two men were thought to be the first Spaniards to set foot on Mexican soil.

Today on the beach at Akumal, where CEDAM has its world headquarters, there is a lovely little statue of Guerrero, his wife, and children put up by Pablo. A plaque informs the visitor that this was the first union of American and European blood in the Western Hemisphere, the first such family. Guerrero is known as the Father of the Mestizo race.

After exploring this wreck site, the CEDAM divers moved on to investigate three others. The first of these, they later concluded, was an English merchantman believed to have gone down between 1823 and 1850. For centuries Alacran Reef had been a plague to captains making their way on a direct course between Veracruz and Cuban ports.

From this wreck they recovered some brass candlesticks and immediately dubbed the wreck site Candeleros. Also brought up from the bottom was a large supply of English household merchandise, Wedgewood china, silver-plated sugar tongs, dinner bells, flat irons, brass bedsteads, and metal-framed mirrors.

One curious find was two jars of Halloway's ointment. The label read: "Sold by the proprietors and other medicine vendors throughout the civilized world." When opened, the jars still emitted a faint, sweet aroma.

Another wreck site evidently was that of a steam-powered

side-wheeler. They named this wreck Columnas after the ornate iron columns supporting the heavy engine room components. The third wreck had little to offer, scant remains from what had been a large sailing ship found in a sand-strewn area at the north end of the reef.

When it was over Andy Rechnitzer said one thing was certain—the coming summer CEDAM would be back to continue its excavations of these wrecks while broadening the search for others.

Fame of the organization was growing. The previous year, 1966, Pablo had received a telegram from Coles Phinizy, of *Sports Illustrated,* who was in the Bahamas attending the Sixth Annual Convention of the Underwater Society of America. His telegram of congratulations told Pablo that he had been awarded one of the coveted Nogi Awards for 1965.

These awards were presented by the Nogi Foundation to individuals who had shown outstanding achievements in the various fields relating to the conquest of inner space in the previous year. For 1965 the awards had gone to Dr. Eugenie Clark, the shark expert, honored for the Arts; Edwin A. Link, the inventor, for the Sciences; John Ernst for sports; and Pablo Bush for Administration in organizing the successful meeting of the Underwater Society in Mexico City. Jacques Cousteau won this latter award the following year.

The year 1966 also had sad news for CEDAM members. Alma Reed, their beloved historian, died. Pablo was away from Mexico City when this happened. Returning sometime later he was abashed to find she remained unburied, her ashes reposing in a small wooden box in a funeral home. He tried to claim her remains but was refused because he was not a relative.

Alma's brother in San Francisco released the remains of his sister to CEDAM, and the American Embassy authorized their being turned over to Pablo representing the organization. He was determined she would not fade into obscurity or suffer the ignominy of having her ashes repose in the cold depths of a funeral parlor.

At first Pablo tried to get Alma buried in Chapultapec Park

among the country's leading poets and writers, or, if not there, in the section of the city's cemetery where many famous Mexicans are buried. The authorities refused either site because she was not a Mexican.

The resourceful Pablo then took her ashes in the little wooden box to Merida, where he consulted with his good friend and CEDAM representative, the late Doña Lucila Dias Solis, who had good friends in the state government. Together they went to see the governor.

This top official authorized them to pick out a spot in the city cemetery but cautioned it should be as far away as possible from where Felipe Carrillo Puerto, Alma's fiancé, and his brothers were buried. These men were now regarded as heroes. The reason he did not want the lovers buried close together was rather delicate. Señor Puerto's widow was still alive, and although the governor was well aware that she and Puerto had been divorced and he intended to marry Alma, he didn't want any repercussions.

Pablo and Lucila went to the cemetery with a letter from the governor authorizing them to pick out a plot. They were lucky. They found an empty plot just in front of the big monument that had been erected for the slain Puerto brothers. When they returned to the governor to obtain his signature, he diplomatically put his name on the paper and never inquired the location of the plot.

Alma was buried with great honors. In Merida it was declared a public holiday so all could attend her funeral. It started at the American Consulate. There were thousands of guests, eulogies, fireworks, parades, music, tears, and a fine monument with Mayan figures erected by CEDAM. On it her role as their historian was noted. Many poems were read in her honor. Alma's song, "Peregrina," was played again and again, and a new song composed by local poets, "Peregrina Returns," was introduced.

My wife and I came to know Alma during our time in Mexico City and we became friends. She was an intellectual, a compassionate human being deeply involved with Mexico, its past and its present. It was no wonder Mexicans loved her and grieved over her tragic love affair.

Her song, "Peregrina," "The Pilgrim," remains ever popular and can be heard everywhere today in Mexico. To those who knew Alma, its strains bring back her memory. To those who did not, tourists by the hundreds of thousands, it is still a lovely Mexican song.

Pablo had a bronze bust made for Alma. It stands in CEDAM's museum at Akumal looking out to sea.

14

In the fall of 1967 a caravan of twenty trucks started out from Pompano Beach, Florida, transporting over 100 tons of men and equipment bound for the Yucatan, 3,500 miles away, in one of the largest underwater ventures ever.

The expedition's destination was the Sacred Well of Chichén-Itzá. It had been organized by Norman Scott, who had supervised diving and the airlift operation at this famous *cenote* during CEDAM's 1961 expedition.

Scott's company, formed for this exploration, was called Expeditions Unlimited, Inc. It was financed by Kirk Johnson and involved the loan or donation of supplies and equipment by a variety of Mexican and U.S. manufacturing concerns.

CEDAM was a sponsor and its divers once again would take part in the adventure. The excavation would be under the direction of Mexico's National Institute of History and Anthropology.

After reaching Mexico City the caravan headed down toward Merida and Chichén-Itzá. Just 250 miles from their destination word came of the approach of Hurricane Beulah. The drivers drove the remaining distance in a blinding rain. Beulah hit Cozumel Island a slashing blow, leaving much wreckage and six persons dead. Then it slewed off, missing Merida and heading up the gulf. The drivers joked—was Chac, the rain god, trying to tell them something?

This time the initial assault on the well was an effort to pump it dry and then use conventional dry land archaeological methods to dig out artifacts. Tremendous pumps were set up that could eject 100,000 gallons an hour from the *cenote*. When the level dropped to fifteen feet, the effort was discontinued for no further progress could be made. They had reached the level of the water table of the peninsula. Underwater archaeological techniques would have to be employed. This possibility had been anticipated.

In their earlier excavation the divers had been hampered by the pile of huge stone slabs on the bottom. To combat this the new expedition had brought along a thirty-ton crane. A number of the stone slabs were removed, but not all for fear of damaging artifacts or producing mud slides.

Poor diving visibility had been a major problem in the past, along with INHA's objection that working in the dark like this made it impossible to assess the historical stratigraphy of the recovered artifacts. To meet these problems a new technique was to be tried. They would treat the well like a gigantic swimming pool and filter the filthy water. The filtration system, supplied by the Purex Water Company, pumped nine million gallons of water from the pool, running this through a bank of filters before returning the water to the *cenote*. The result was crystal-clear water.

Everyone was delighted with the outcome of the experiment. For the first time the divers had excellent visibility and underwater photographs could be taken. INHA archaeologist Victor Segovia directed the stratigraphic mapping of the *cenote,* keeping detailed records of where the artifacts were found and in what sequence.

Modified airlifts were used to lift the mud in layers. Any object larger than a button was brought up to the surface by hand. All artifacts were catalogued, tagged, and sealed in water before being sent on to Mexico City for further investigation.

Artifacts recovered included two beautifully carved wooden stools; several wooden buckets; about 100 jars and vases of different sizes, designs, and periods, some almost complete and elaborately painted; fabrics; gold items; rope; rings; bells; flint arrowheads and spearheads; obsidian knives; objects made of jade, rock crystal, copal incense, rubber, coral; human teeth perforated for necklaces; conches; ceramics; bone; mother-of-pearl; horn; amber; copper; quartz; pyrite; onyx; human and animal bones; grinding stones; five stone jaguars; and two stone serpents.

The diving excavation lasted two and a half months, running into 1968, before coming to a close. It would take months or years for the experts to evaluate the artifacts. INHA specialists said one interesting conclusion was it appeared, judging from the human bones found, that many more children than adults had been sacrificed. The ratio was more than 50 percent children. Some were young boys, their heads deliberately deformed from infancy.

As for underwater archaeology valuable lessons had been learned. These were: it would be costly, if not impossible, to drain the Sacred Well of Chichén-Itzá; *cenote* water could be cleared and purified, thus aiding diving investigations; and an airlift, properly handled and controlled, could be used to dig underwater sites in *cenote*s, making it possible to get the proper stratigraphy.

When it was over, Pablo Bush once again said it remained his opinion that only a slight amount of the artifacts of the Sacred Well had been recovered. This expedition had been better than the last. The next would be even better.

In 1967 the famous Boston Sea Rovers held their thirteenth Annual International Underwater Clinic at the Massachusetts Institute of Technology. Proof that CEDAM International was getting ahead in the world was the number of experts in attendance who had become members, people like the fabulous Dr. Harold E. "Doc" Edgerton, of MIT, whose pioneering research had been the foundation for the development of the electronic speed flash; Peter R. Gimbel, the department store heir, known for his many underwater trips to the wreck of the *Andrea Doria* and his daring film footage on sharks; Jon M. Lindbergh, who piloted the submersible that led to the recovery of the lost H-bomb off Palomares, Spain; Coles Phinizy, senior editor of *Sports Illustrated,* 1964 NOGI Award winner for art;

Paul J. Tzimoulis, editor and publisher of *Skin Diver*; M. Scott Carpenter, astronaut and aquanaut; Dr. Eugenie Clark, leading shark expert, and many others.

Diver of the Year Award was presented to Edwin A. Link, inventor, explorer, hydro-archaeologist, a CEDAM member who had taken part in the earliest expeditions.

Before this distinguished company Pablo Bush was given a Sea Rover's Special Award with the citation: "Diving statesman, founder of CEDAM, steadfast leader who enlarges diving horizons wherever he goes."

Regarding Pablo and this event, the Sea Rovers' booklet said: "Anyone reading about diving in the Caribbean notices his name time after time as the moving spirit in that large area. This multifaceted character was the force which united the few divers with many interested and powerful people in Mexico to form CEDAM, an organization that has Mexican government recognition."

15

In 1968 CEDAM International began to spread out in a variety of directions. Early in the year one investigation was on the west coast of Mexico south of Puerto Vallarta, where CEDAM's long-time photographer, Genaro Hurtado, had happened upon an old anchor in shallow water with a few links of chain still attached.

Could this be from one of the ships of Hernando Cortes, the Spanish conquistador, believed to have sunk along the west coast of Mexico between 1529 and 1539? A small CEDAM International expedition searched the area for five days of tough diving in heavy surge and found nothing.

This expedition was partially financed by the chairman of CEDAM International's board of directors, a colorful Texan by the name of Carroll H. Shelby, a top international racing car driver, designer, and builder. On vacation in Quintana Roo, he had met Andy Rechnitzer, who persuaded Shelby, who was a poor swimmer and had a badly swollen sprained ankle, to don scuba gear and learn what it was all about.

Holding his cane with one hand and a tether with the other, Shelby was towed around underwater until he reached the thirty-foot contour. He became a dedicated diving enthusiast, so dedicated, in fact, he established a city commissioner post for "Oceanography" in his privately owned, landlocked city of Terlingua, Texas, "the Chili Capitol of the World."

In the summer of 1968, once more CEDAM International and CEDAM of Mexico teamed up to continue their investigation of wrecks on Alacran (Scorpion) Reef. The Mexican group took the first two-week period, the Americans the second, visiting nine wrecks during the thirty-day period.

The divers were transported from the little port of Progreso to the reef aboard a fifty-two-foot trawler, the *Yoyis*. Andy had with him his wife and scuba-diving children, David, Andrea, Martin, and Michael. Also along was Joe Kelly Hughes, a former U.S. Navy Seal, later to become CEDAM's director of operations at Akumal. Paul Tzimoulis, editor of *Skin Diver,* took part, planning a major article on the exploration as well as another movie.

Their efforts were making the reef famous. A documentary motion picture entitled *The Treasure of Scorpion Reef,* shot by Genaro Hurtado the previous year, had already been viewed by thousands of people. It had won "best of show" at the Eleventh International Underwater Film Festival as well as a gold medal at the Eighth International Underwater Photo Exhibit.

Revisiting the Candeleros wreck they could see that other divers had been there before them, pilfering the site of larger brass items, such as bedposts. Nevertheless, they were able to pick up a number of smaller objects that would possibly help them in their follow-up efforts at identification.

They also returned to the *Columnas* wreck, where they made extensive measurements of the steam engine. The divers were now presuming this ship to have been a side-wheeler of about 1850. At the site they found several coins, one of silver with the Mexican eagle and the date 1845, another a Spanish gold piece dated 1790.

A unique first was the discovery at this site of mercury. Dr. Ted Sharp, swimming among the coral ravines, sighted two small beads of mercury caught in the algae. Joe Kelly Hughes later discovered buried in the coral five flasks that contained nearly pure mercury.

A new wreck site was located on the north end of the reef. Along with several cannon balls, three cannon bearing the crest of King George III were brought up and transferred to Progreso for preservation. Each of these weapons bore distinctive serial numbers, the broad arrow, the date of manufacture, and the manufacturer's insignia.

Nearby the CEDAM divers found evidence of still another wreck where an abundance of cobble ballast stones were strewn over a large area surrounding a mass of fused iron rods with several coral-encrusted cannon and an anchor. It was thought these two ships were related and were of the eighteenth century.

Always leap-frogging, thinking of future expeditions, Pablo Bush, Tom Fifield, a CEDAM International member of the board of directors representing the Milwaukee Public Museum, Dr. Stephan Borhegyi, an expert in underwater archaeology from the same institution, and others investigated the Chinkultic *cenote* near the Guatemalan border of Mexico. They concluded this would be a prime target for future exploration.

In the fall at an annual meeting of the British Sub-Aqua Club in Brighton, England, CEDAM International members made a major contribution with their lectures: President Rechnitzer on "The Status and Operational Accomplishments of U.S. Deep Submersibles," Norman Scott on "The Methods of Underwater Recovery at the Sacred Well of Chichén-Itzá," and Pablo on "What's New in Underwater Archaeology in Mexico." The British club's founder and others joined CEDAM International, planning to take part in upcoming expeditions.

The trio visited the Royal Artillery Museum in Woolwich, the Wedgewood Company, Goldsmith Hall in London, and the Naval Museum in La Spezia, Italy, on this trip, but were able to add little data to their search for identification of the Candeleros wreck or the one they call the Warship of King George III.

Although CEDAM's thrust was mainly toward unraveling mysteries underwater, in the sparsely populated Quintana Roo area little could happen of a scientific or archaeological nature without the organization in some way getting involved. So it was with the American Quintana Roo expedition's scientific crossing of the territory in 1968.

The expedition was led by Robert O. Lee of Portland, Oregon, an American businessman with a bent for scientific exploration.

Scientists they had aplenty. What was needed was someone to accompany them who could speak Maya. Pablo was contacted. He put the group in touch with Ramon Mendoza, CEDAM representative in Valladolid, close to the town of Chemax, in Yucatan, the starting point.

When Lee learned of the work of CEDAM, he immediately joined the organization. Later he became a member of CEDAM's board and of its expeditions committee.

Departing from Chemax the American Quintana Roo expedition eventually covered 150 miles, 114 of which was through completely unexplored jungle. It was the first scientific crossing of this wild area, the team walking from the inland town of Chemax to the Caribbean coast.

The expedition, one of the most extensive in the history of the Yucatan, was composed of a number of specialists. As they moved through the jungle mapping their route, they made observations of villages, trails, water availability, and general ecological conditions in the chiclero territory. They collected 170 botanicals, including 23 species of orchids in the semiarid jungle where walking was made difficult by the rough limestone rock underfoot.

Indicative of the variety of animal and reptile life in the jungle was their collection, which included the fer-de-lance, constrictors, green tree vipers, coral snakes, howler monkeys, raccoons, opossums, foxes, squirrels, skunks, deer, peccaries, iguanas, helmeted lizards, sea turtles, and manatees.

Over 80 bird species were identified; 2,000 insect specimens were collected. Sick Indians they met along the way were treated, and medical studies were made of the health of expedition members under the stress of heat, humidity, and great physical exertion. Archaeological observations were made of Mayan *sacbe*s (stone highways) and other ruins of this ancient civilization they encountered along the way.

All of these happenings were reported to far-flung members in the new quarterly CEDAM International Bulletin, which now made its appearance. In the U.S., it was reported, three chapters had come into existence, at Long Beach, El Paso, and San Antonio, with a fourth forming in Washington, D.C.

The Southern California Chapter at Long Beach, sparked by Andy Rechnitzer, had ambitious plans to affiliate itself with the Museum of the Sea aboard the *Queen Mary*. An exhibit on the accomplishments of the organization was planned. Andy's goal was to see CEDAM International chapters each work on a formal basis with a major museum in the area, providing lay divers to work under the direction of professional nautical archaeologists on serious projects.

By the end of 1968 CEDAM International and its activities were becoming widely known in the U.S., especially among divers and those interested in exploration in a popular sense. *Triton,* on its cover, had a photo of Pablo arriving back to the surface of the Sacred Well of Chichén-Itzá holding up a human skull he had recovered. *Skin Diver* had a lead article on the Alacran Reef explorations. Other articles appeared in popular periodicals like *Dive*, *Argosy,* and *Look.*

On the scientific side was publication of a paper on the exploration of the Sacred Well of Chichén-Itzá by Mexico's National Institute of Anthropology and History.

As President Rechnitzer noted in his report, CEDAM International was clearly a viable organization by the end of its second year.

16

The chain went like this: two tiny beads of mercury had been found in a coral ravine on the 1968 CEDAM expedition to Alacran Reef; this led to the finding of five flasks of mercury; old coins helped establish general dates.

Using the dates as a reference, President Rechnitzer checked an account published in 1892 of early steam-driven merchantmen belonging to Great Britain and concluded the wreck they had named *Columnas* was that of a ship called the *Forth.*

CEDAM International advisor Bob Marx, a leading expert on wreck sites, said no; because of the mercury it had to be a ship called the *Tweed.* A telegram to Lloyds of London brought a quick response. Two royal mail steam packet vessels had been lost on Alacran Reef, the *Tweed* and the *Forth.* From the *Tweed,* in 1847, out of a cargo of 1,115 flasks, divers had recovered over 248 flasks of mercury.

This was followed with research by CEDAM International member Chris Talbert, who uncovered a book in the Clemson University Library entitled *Notable Shipwrecks* by William Senior, published in 1873, giving details of these disasters.

Thus the CEDAM investigators finally knew what they had found.

The remarkable thing was that these two top-notch steam vessels, both engaged in the West Indian mail service, both belonging to the same company, both named after two of the largest rivers in Scotland, should within three years both be destroyed on the same reef.

The *Tweed* was one of the royal mail's first fleet of fourteen paddle steamers. Completed in 1841, she had a hull of wood, three masts, barkentine rigged, and had a gross tonnage of 1,800. The *Tweed* could accommodate up to 200 passengers, a crew of 89, besides carrying general cargo, specie, and mail. Her only armaments were pistols, cutlasses, and a small signal cannon. The *Forth* was similar, but larger by 100 tons.

R.M.S. *Tweed,* commanded by a Captain Parsons, left Southampton on December 17, 1846. After calling among the West Indian islands, on February 9, 1847, she left Havana for Veracruz, Mexico, having on board 62 passengers and a crew of 89. The quicksilver she carried, an excellent solvent, was for use in the silver mines of Mexico. Spilled mercury had gotten on some of the silver coins found by the CEDAM International divers, eroding their surface.

The weather was bad, overcast, with heavy rain and a strong wind. There was no opportunity for taking sights to check the ship's position. Captain Parsons used dead reckoning, steering a course designed to take him midway between the coast of Yucatan and Alacran Reef. The *Tweed* was traveling under both steam and sail at a speed of seven knots when at 3:30 A.M., on the morning of February 12, the cry from the lookout was heard: "Breakers ahead!"

Although the order went out instantly to reverse the engines and put the helm hard to starboard, the *Tweed* struck the reef with a mighty crash. The ship's bottom was dashed. The engines stopped and a cloud of steam rose over the doomed vessel.

Waves threw the ship repeatedly on the rocks, forcing it over on the starboard side, sweeping away three of the lifeboats, bringing down spars, and wrenching away one of the paddle boxes. Many of the passengers rushing to the deck were swept overboard.

In the darkness Captain Parsons directed the lowering of the two remaining boats. These capsized almost at once, the waves washing both survivors and those who drowned on to the reef. In a short space of time the *Tweed* broke in half, leaving "a confused mass of broken timbers, splintered planks, cordage, tangled spars, and human beings." In a way this was a godsend for many people were able to save themselves; they clung to pieces of debris. Captain Parsons had been thrown into the sea when the stern went down; though injured, he saved himself by hanging on to a spar.

Alacran Reef was entirely submerged. At dawn the survivors, standing in a foot of water on the jagged coral, could see only ocean from horizon to horizon. On the edge of the reef they could count forty persons still clinging to the bowsprit and forward part of the ship as the waves dashed over them. During that day and night those trapped on the derelict were drowned.

Fragments of the wreck extended for a mile along the reef: "Cabin doors, pieces of boat, casks, baggage, seats, furniture, beds, seamen's chests, and passengers' luggage formed one continuous ridge, in places several feet in height."

The first fear of the battered survivors was that they would be overwhelmed by the tide, a fear put to rest by Captain Parsons, who assured them it would not rise above three feet. He organized the castaways, crew and passengers alike, into three working parties to salvage anything that might be of use from the wreckage. All but the injured were set to scavenging.

Their "loot" was varied: casks of flour, oatmeal, butter, brandy, cases of wine, three pigs—two alive—a live sheep, a dog, clothing, vinegar, candles, matches in a waterproof container, and one of the ship's compasses.

For the moment the horrible possibility of starvation could be put off. Vinegar, a substitute for water, was doled out in sips.

Next the captain ordered construction of a platform from floating pieces of debris, lashing these to spurs of coral. At first only ten feet square, the platform was a place to store provisions and where survivors could take turns trying to sleep.

While this was going on, others were at work trying to repair a mailboat that had been recovered. Nails were found, a copper bolt served as a hammer, odd sails were picked up, the compass and some of the provisions put aboard. Ten men were selected to man this rickety boat, the ship's chief officer being in charge. Six sailors took the oars. The American vice-consul at Veracruz was sent because he spoke Spanish.

By four o'clock in the afternoon, following both prayers and cheers, the frail boat was off to traverse seventy miles of ocean to the Campeche coast. As the boat disappeared over the horizon, the survivors had their first meal, a ball of oatmeal mixed with flour and salt water and a bit of wine to drink.

The wreck had occurred in the early hours of Friday morning. Under Captain Parsons' direction, by Saturday afternoon the platform's size had been doubled and strengthened. A spar flying a signal was in place. Scavenging continued and skills were sorted out as the 69 persons on the reef waited for deliverance.

By Sunday, using a sheet of iron on the platform and dried wood, a fire had been built, the dead sheep cooked along with some fish and lobsters they had caught. Using the lid of a metal trunk, cakes had been baked. They even found a way to distill a little fresh water.

On Monday, the fourth day, construction was begun of a large raft that could be used, in extremity, to try to reach the coast. While this work was going forward, a sail was sighted on the horizon! It was a brig towing a canoe. This latter craft was manned by natives. As it drew nearer, while the brig lay off at a safe distance, they could see their chief officer standing in the bow waving his hat at them.

The chief officer told the enthusiastic castaways that, with two men bailing constantly, they had made good progress, and by Sunday had met the Spanish brig *Emilio,* six miles off the mainland. Once assured the men in the mailboat weren't pirates, the brig had put in to the small port of Sisal, hired the canoe, and sped to the rescue.

Using the canoe as a ferry was slow going with bad weather and rough surf. Several times it almost capsized. Before night fell only 26 people had been taken out to the brig. The remaining castaways would have to spend another night on the reef. To prove he wouldn't desert them, the captain of the brig, Don Bernardino Campo, passed the night with them.

On Tuesday the canoe returned along with the brig's small boat. Those left were loaded aboard. Captain Parsons was the last to leave the reef. Sounding with poles and rowing with great caution they tried, because of the foul weather, to reach the small islet of Perez where they would board the brig rather than try the open sea.

By nightfall they had made only four miles and had to anchor, spending another sleepless night on the reef. Next morning all were finally aboard the rescue ship. At Sisal their injuries were treated. From here, again on the brig *Emilio,* they were taken to Havana and from there, on R.M.S. *Avon,* carried back to England.

Thus, the story of the *Tweed.* Some 72 persons perished in the wreck. Both Captain Parsons and the Spanish captain were later given medals and monetary rewards.

The wreck CEDAM International had called *Columnas* had become the *Tweed* and the details of her demise were now known. President Rechnitzer said these remains were an excellent example of an underwater wreck that should remain untouched for posterity. The rusting iron and corroding copper should be left, he said, as a source of visual enjoyment—an underwater monument—for divers swimming about a wreck that had gone down more than a century ago.

Not long after CEDAM's exploration of this wreck some of its divers returned to the scene. They found someone had blown *Columnas* to bits.

So much for the underwater monument.

17

One way to look for shipwrecks is to search the literature, establish that a certain vessel has gone down and the specific location, and then go diving to look for it. The other way is to look around underwater, find a wreck site, then through research try to identify the ship. Using this latter method CEDAM International had found out about the *Tweed*. By chance they also had information, based on the first method, about her sister ship, the *Forth*.

R.M.S. *Forth* was coming from Jamaica when, on the night of January 13, 1849, she struck Alacran Reef, almost exactly three years after the *Tweed*'s disaster. Lieutenant Molesworth, a giant of a man who was to be the hero of this mishap, was in his bunk when it happened.

He heard the ship bump something, then an awful crash. He knew at once they were on the reef! This was followed by blow after blow. On deck passengers were rushing about as the ship swung from side to side, making it almost impossible to stand.

A quick survey revealed the *Forth* was stuck on a projection of the reef that had come through the bottom of the steamer into the engine room. The engines had stopped. Sea water had put out the fires. She was dead on the reef, the surf lifting her, then throwing her back on the rocks.

The captain ordered Lieutenant Molesworth to get the women and children into the largest lifeboat. The problem was that it was in the water twenty feet below the bulwarks. He made fast a rope, took the first lady around the waist with one arm, and with the other lowered himself down. In this way he put into the boat all the women, four children, and a stout old gentleman almost out of his mind with fear.

Out to sea someone sighted a small vessel, but it was drawing away from the reef. Lieutenant Molesworth volunteered to go after it. With four volunteer sailors in a small boat they cleared the breakers and began to row. Within an hour they had caught up with the "small vessel," which turned out to be three small sailing canoes. They refused to come close to the treacherous reef, but were willing to go for help. The lieutenant agreed he would try to get everyone to Isla Perez, about eight miles from the wreck, to await rescue. This was the same islet where survivors of the *Tweed* had been picked up.

Following this plan, all boats were lowered as the *Forth* twisted and heaved on the reef, and a course was set for the islet on the southwest side. This time they were able to load the boats with supplies.

Landing on Isla Perez, they found two dilapidated shelters for fishermen, but, fortunately, they were stocked with six barrels of fresh water. The islet was about a mile in circumference. Again the castaways feared starvation if help did not arrive. They were put on short rations. But luck was with them. The next day the Spanish brigantine *Isabella* turned up, which they immediately "chartered" to take them to the Mexican mainland. But first the valiant Lieutenant Molesworth, in one of the brigantine's canoes, along with a Spanish sailor crossed the reef thirteen times, returning to the doomed vessel to recover passengers' baggage and other valuables. This time there were no fatalities.

It was natural during the 1969 expedition of CEDAM of Mexico and CEDAM International that they keep a sharp eye out for the remains of the *Forth*. The expedition even based itself on Isla Perez. Excitement rose when they thought they had found her, something on the bottom that looked like the spokes for a side-wheeler.

Closer examination revealed this to be part of a narrow-gauge railway, complete with tracks, encrusted in the coral. There were seven pairs of locomotive wheels and a piece of corroded sheet metal, once the shell of a boiler. Among all this debris no ship remains could be detected. Another mystery. The divers concluded perhaps these items had been part of a deck load on a ship caught on the reef, freed after this dead weight was pushed over the side.

Once again the expedition was divided into two phases. Group one during the first two weeks was largely Mexican and led by Enrique (Gitano or Gypsy) Enriquez. It had the task of exploring and establishing the condition of wreck sites. Group two during the second two weeks was mainly an American effort, although there were overlaps. It was led by Andy Rechnitzer. They would continue exploration and excavation and film a documentary movie.

Night in these tropic areas can be magical, especially when the moon is full. This time group one camped on Isla Perez. The moon hung as a great golden ball in the sky. They were able to watch both it and, because of the unanticipated wonder of a fisherman having a TV set on the island, also America's history-making landing on that heavenly sphere.

But during the first phase on Alacran Reef things were not going so well. The air compressor continued to lose capacity despite all efforts to fix it. This hampered operations. Diving is hungry work, and the cook turned out to be incompetent, obstinate, and temperamental. This did not help. And their trip back across the water to Progreso was rough, wet, and exhausting. Nonetheless, they completed their assignment.

These problems were rapidly straightened out. Andy and the others arrived for phase two. A new 3.5 cfm compressor, a gift from the U.S. Divers Company, was air-freighted to Cozumel Island, where it was picked up by the *Don Luis A.,* the 65-foot boat the group had chartered. Gypsy Enriquez agreed to stay on as a co-captain and do the cooking along with his leadership responsibilities.

Everything fell in place. Everyone was given a number as an easy means to check on the return of divers. Some went out for food, fish, conch, lobster, turtle, and octopus. Gypsy turned out to be a gourmet cook. Others pursued their tasks of documenting on both movie and still film this underwater archaeological "dig." Research had made them more knowledgeable.

The wreck they had called King George III or *Pecio Hernan,* through correspondence with authorities in England, was believed to be the remains of the British man-o-war *Tay,* sunk November 11, 1816.

Her commanding officer, Capt. Samuel Roberts, had written to his headquarters, saying, "I lament much that I have to acquaint you, for the information of the Admiralty, of the unfortunate loss of His Majesty's Ship, *Tay,* under my command, being wrecked whilst beating up from Campechy to Havannah."

He had been stung by Scorpion Reef. At first he tried cutting away his fore and main masts, hoping to use these as a bridge to get the 132 persons aboard, passengers and crew, over to the reef. This plan failed when the masts broke in their fall. Then the captain ordered everyone into the boats.

Working their way through the shallow water, they managed to get to the islet of Desterrada on the northeastern end of the reef. From there he sent one of the small boats to the mainland for help.

Captain Roberts had brought with him to the islet specie worth $730,000 being carried aboard his ship. All those camping in the open on Desterrada for a week were relieved when a Spanish corvette appeared. Somehow these pirates learned about the treasure, turned their guns on the group, and sailed away with it. The castaways had to wait until later to be rescued.

The divers returned to the site of the Candeleros wreck, so named after the candlesticks they had found there on the sea bottom. Research had now indicated this was the wreck of an English merchantman, *Halliday,* skippered by a Captain Steward. It had gone on the reef on May 6, 1822. All aboard were saved.

The expeditionaries had collected a representative sample of her manifest.

At another site cannon, ballast stone, musket balls, and glass stoppers were excavated. They named this site *Pecio Nuevo.* Later it was thought possibly here lay the bones of the *Sancho Panza,* which went aground on Scorpion Reef in 1824.

Not having sophisticated equipment for searching underwater, one CEDAM approach was to buy beer for fishermen in the local bars, gain their confidence, and ask if they knew of any wrecks. A number had been located in this way.

Gypsy Enriquez had been drawn by the tale of a diving lobster fisherman who claimed to have seen a large cache of artifacts on the bottom and had brought one back as proof. It was a piece of chinaware bearing a Bordeaux, France, mark of origin. On a return trip the fisherman was unable to locate the place. Perhaps the sands had shifted.

On the basis of the piece of china and an examination of CEDAM advisor Bob Marx's unpublished "List of Ships Sunk in Mexican Waters (1816-1824)," it was thought to have come from the wreck of the French ship *Cacique,* en route from Bordeaux to Veracruz. In 1824 it caught on the reef and was a total loss. The years covered in Marx's publication seemed to have been an eight-year period of disaster for ships trying to skirt Alacran Reef.

Fishermen can weave some wonderful tales over beer. Both they and the divers would become excited. Who can sort fact from fancy? Better to check it out, not take any chances. So on this expedition, as in the past, once again time was spent cruising deep waters hoping to find the elusive *Black Dragon.*

This ship was said by the Mexican fishermen to be fully intact, resting upright on the bottom with its conspicuous black bowsprit, carved in the form of a dragon, clearly visible through the water. Nothing was seen. Could it be true? Wouldn't it be wonderful if it were?

Once again the expedition had its gleanings from the sea: English and Mexican gold and silver coins, cannon balls, muskets, a sword handle, a fork and a silver platter, along with other items. All were listed with the Mexican secretary of Patrimonio Nacional and placed in CEDAM's new museum, which, true to his word, Pablo had opened in Mexico City.

Meanwhile, CEDAM International had continued to grow. Three new chapters, in Washington, D.C., Dallas, and New Orleans, had come into existence. Earlier in the year the organization's second annual meeting had been held at Santa Monica. Those attending saw a new film, *Coral Archive of Alacran Reef,* by John C. Mahon, a Hollywood producer, based on the previous year's expedition. There also was a new film, *Black Coral,* by photographer Genaro Hurtado.

Four individuals who had made outstanding contributions to underwater archaeology were elected as honorary members of CEDAM International. These were Jacques Yves Cousteau, Dr. Ignacio Bernal, Dr. George F. Bass, and Luis Marden.

Dr. Bernal was director general of Mexico's National Institute of Anthropology and History. He had given his personal support and encouragement to the organization's Chichén-Itzá excavation. Dr. Bass was recognized internationally as a leader in the field of underwater archaeology. Marden was chief of the foreign editorial staff of *National Geographic* and a distinguished underwater photographer.

Reflecting Andy Rechnitzer's goal of bringing chapters closer to area museums to work on serious projects was a talk given at the meeting by Dr. Charles Roszier, curator of archaeology, Los Angeles County Museum of Natural History. He was an expert on Indian relics found on the Channel Islands of southern California and had been doing dry land digs for artifacts there for years.

He was eager to point out the need for underwater archaeological exploration of the island waters and was looking forward to members from the Southern California Chapter of CEDAM International to help with this offshore diving.

Between expeditions or local diving CEDAM International members could now increasingly have a vicarious share in the adventures of other members. Bob Marx was planning to sail an authentic reproduction of a Viking ship from Denmark to the New World.

Peter Gimbel, an outstanding diver who had gained fame with his photographs of the *Andrea Doria,* was off to the Indian Ocean to make an underwater film on the great white shark. This later became the dramatic feature film, *Blue Water, White Death.*

Some CEDAM International members were even crazy enough to want to go on an expedition like this, if they could, instead of digging around for Indian relics. The organization and the ocean were big enough for both.

18

For 1970, CEDAM International announced four expeditions, the first to take place in the spring with a survey by balloon of the Quintana Roo reefs to be led by Rechnitzer, Don Piccard, and his cousin Jacques Piccard of Switzerland. The idea was that from a balloon, towed by a boat, they could look down into the clear water and see objects on the bottom. For various reasons this fascinating project had to be scratched and has never taken place.

In midsummer three expeditions were planned, again in concert with CEDAM of Mexico. One would be to Alacran Reef, led by Rechnitzer and Enrique "Gitano" Enriquez. At the same time there would be another, led by Alfonso Romero—now president of the Mexican dive group—to Chinchorro Reef, and a third, less strenuous, from Akumal to Xelhá, led by Pablo Bush.

Actually there was one more—the Chinkultic Area Archaeological Exploration and Excavation Project begun in January. Preliminary investigations of the *cenote* there had been made by Pablo, Tom Fifield, and Dr. Stephan Borhegyi, director of the Milwaukee Public Museum in both 1968 and 1969. Work had been delayed on the death of Dr. Borhegyi the previous fall.

To get to Chinkultic one could drive the whole length of Mexico heading for the small town of Comitán in the State of Chiapas close to the border of Guatemala, or fly to a small airstrip at Comitán.

Chinkultic, which means "place of the caves," was situated at 4,500 feet in the rugged highlands surrounded by the Lakes of Color (Lagos de Colores) and the Lake of Montebello. After scouting the area from the air Pablo wrote: "These lakes are some of the most beautiful and incredible I have seen in the whole world. They look like necklaces made of precious jewels of all hues of blues and greens."

Pablo was not ready to give up on the idea of pumping a *cenote* dry, even though this technique had failed at Chichén-Itzá. He had organized the expedition on behalf of CEDAM of Mexico, drawing once again on both his American and Mexican contacts for support.

An American, Robert Soehnlen, president and owner of the Beloit Pipeline Construction Company, put up the major financing as well as his know-how and machinery for pumping out the well. Archaeologist Roberto Gallegos was appointed by Mexico's Institute of History and Anthropology to head the project with two assistants, one to make paintings and drawings. Tom Fifield and his wife, Marilyn, along with Lon Mericle, represented the Milwaukee Public Museum, which had agreed to finance Mexican laborers on the site. Andy Rechnitzer went down, but this time there was no diving.

A great many Mayan archaeological ruins in both Mexico and Guatemala are covered by jungle. Chinkultic was no different. The first job was to get rid of trees and underbrush. Gallegos offered to supervise this work. Fifield placed the laborers under his direction. CEDAM paid for needed tools and materials. They were also able to get some of the citizens of Comitán to donate construction materials.

As the jungle was cleared many archaeological sites came to light. One was a 200-foot staircase leading up the side of the mountain to the *cenote,* where there were four platforms and two large pyramidal structures. This made it probably the highest pre-Colombian structure in Mexico, almost as high as Tikal, the tallest pyramid in Guatemala. Gallegos also turned up a ball court near the well and found sixteen stelae lying under heavy brush.

It took until March to get the site cleared and the pumps installed. Then, by pumping night and day, they managed to bring down the water level 18½ feet, exposing about 80 percent of the bottom of the funnel-shaped *cenote*. Though they were working in wet mud, the men would use a dry-land technique rather than diving.

Bob Soehnlen, calculating the budget and daily cost, told them he could keep the well dry for only ten days. But on seeing the dismay of the archaeologists, he upped this to a maximum of twenty days during which they would have to do their work.

Gallegos had eleven exploratory trenches dug, then concentrated on the most promising. The result was thousands of potsherds. These were turned over to the New World Archaeological Foundation for study at their workshop and museum located in Tuxla Gutierrez, capital of Chiapas. A clay idol of Chac, god of rain, was recovered, the first from a *cenote*. Also among the many artifacts were grotesque ceramic figures. The expedition proved a *cenote* could be pumped dry and archaeological work accomplished.

The four platforms at the edge of the well contained tombs. These had been robbed. Neither in the ruins nor in the well was anything of monetary value discovered—no jade, gold, or metal of any kind. Nor did they find any human bones. Evidently no human sacrifices had been made here.

The group also explored a number of surrounding caves. Many had stairs leading down into the interior. Precautions were taken because a year earlier several investigators had become seriously sick from breathing the dust from bat dung.

When they left there was the tantalizing thought. If only we could have had the time to go deeper in the well, or explore more caves, or do scuba diving in the lakes—who knows?

In February 1970, CEDAM International held its third annual meeting at the Los Angeles County Natural History Museum. This was timed to coincide with the International Underwater Film Festival. Their new film, taken under Rechnitzer's direction by Bob Dingman during the 1969 Alacran expedition, was entitled *Five Fathoms to a Ship's Grave,* and it showed search, identification, and excavation of a wreck site. It won the "best of show" award.

Pablo Bush briefed the 450 members who turned up for the meeting on the Chinkultic expedition. Based on his leadership in underwater exploration, he had just been given *Argosy*'s Giant of Adventure Award for 1969. The magazine's story about Pablo's adventures was written by Robert F. Marx, who had won a similar award for sailing the Viking ship across the Atlantic. From a worldwide group of candidates Pablo had also been given the Gold Plate Award of the American Academy of Achievement at Dallas, Texas.

The annual meeting was better than ever, embracing a comprehensive series of lectures as an indication of the organization's broadening horizons. Dr. Sol Heinemann, M.D., chairman of CEDAM International's board of directors, told of a special study he had made of health problems and preventative steps for underwater explorers in the salt and fresh waters of tropical Mexico.

Norman Scott, of Chichén-Itzá fame, told of his current efforts to raise the famous ship *Tecumseh* from the mud of Mobile Bay for the Smithsonian Institution. Other authorities spoke of research and excavations of historic ships from California to Greece. Of special interest was a talk by Louis Nackos, head of Mayan Tours and Expeditions, who each year took a group of novice explorers on rubber rafts along a portion of the coast of Quintana Roo. He had been tantalized by frequent sightings of small temples with doors and rooms suitable only for a dwarf. Were they built like this to force worshipers to crawl in on their knees? Honorary director Paul Tzimoulis, editor of *Skin Diver,* spoke optimistically of the "New Horizons for Underwater Archaeology."

Besides the museum in Mexico City, CEDAM had opened another small museum and support facility at the port of Progreso. President Rechnitzer said it was an expression of confidence in the organization that the Mexican government had agreed that all of the Alacran artifacts could be kept for display in these two museums.

In June, Andy put into practice what he had been preaching. Along with N. Nelson Leonard, chief archaeologist of UCLA's

Archaeological Survey, he led thirty-five members of the southern California chapter of CEDAM International on an expedition to Catalina Island.

Artifacts previously recovered from the island, twenty-seven miles off the coast of southern California, indicated human inhabitants for at least 4,000 years. To make an underwater exploration off Little Harbor on the seaward side of the island, the group broke down into small parties. No artifacts were found in the study area.

Undismayed the divers put aside their scuba equipment, picked up shovels, and under Leonard's direction at Big Springs made a total of seven one-by-one meter pits. From these, 75 tools, 89 waste flakes, and approximately 900 pieces of bone were recovered. Their research generated new questions. Leonard was convinced of the value of cooperation between groups like CEDAM and professional archaeologists. He wanted them to do more underwater exploration in the future.

While this was going on in California, another miniexpedition was taking place in Mexico where two Texas housewives, Janis M. Keller and Julie Lama, traveled to San Cristobal de Las Casas, in the State of Chiapas, to meet an American woman, Trudi Blom, a member of CEDAM who was to be their guide.

Writing of this later, Janis Keller described how they floated in a mahogany dugout canoe down the Usumicinta River saying, "In the wonderful world of CEDAM International adventure is practically a way of life."

Trudi, with her husband, Frans, had gone to live in the area in the 1950s to study Meso-American archaeology and ethnology. Her husband had died in 1963 and she was carrying on the work. In this she had come to have a special relationship with the Lacandonian Indians, Mayans living in the rain forest, their aquiline profiles the same as those seen on ancient Mayan stelae.

The Lacandonians had never been converted to Christianity. Women were not permitted to enter the "god house," but in this case, in a ceremony to benefit Trudi's health, the three women were permitted to sit in the rear of the god house and watch while copal incense was burned. A palm fan was waved through the smoke and tapped lightly on Trudi's head.

Also beginning in June—still another first—CEDAM International teamed up with the Fort Worth Museum of Science and History, with Jerry A. Griffin, under the supervision of INHA, bringing together qualified museum and college personnel for six weeks of mapping and photography in the Akumal-Xelhá area.

As usual the centerpiece of the year was the midsummer Alacran Expedition with the Mexicans taking the first two weeks and the Americans the second. Language was a barrier for some to make a total integrated effort, plus the inability of most participants to get a month off from work.

Each group was restricted to twenty persons. The Americans came from all over the United States. They were advised to bring their own tents, cots, or hammocks. They were told to bring a blanket and mosquito repellent, this for the mainland, not the reef, where there were no mosquitos. They needed medication for sunburn and scratches, protective, lightweight clothes for both above and under the water, gloves, a hammer and a chisel, a plastic screen bag for artifacts, and tennis shoes. They would need a plastic plate, cup, and bowl personally marked. Spearguns were welcome—the food was needed.

They were to be prepared to accept chores, these being shared by all participants. Once embarked it would not be possible to return to the mainland except for an emergency. Diving would be rigidly monitored. And, above all, they were urged to come in good health. On each expedition there were some old-timers and some newcomers.

During phase one the divers were led to a new wreck site by the captain of the boat they had hired in Progreso. After an investigation they concluded that here were several ship strewn one on top of the other. Perhaps they had located the remains of the *Forth*. They visited the *Tweed* site again and brought up a small cache of silver-plated flatware, the cover and works of a pocketwatch, a gold bar from some spectacles, a bit of gold watch chain, and a small number of coins.

Phase two was not what some might call ideal. The largely American group traveled over rough waters in a small boat that anchored for the balance of the night well out from Isla Perez. By dawn almost everyone was seasick, some violently. Almost all the supply of seasick pills and injections had been used up.

Once ashore many were too weak to work, most couldn't eat, and others put up the kitchen and dining area and made camp. The weather remained foul; high winds and surf made diving outside the reef impossible. They were confined to "joy" dives inside the reef.

Gary MacFadden, swimming in a sheltered part of the reef just before Hurricane Celia put an end to the whole business, found a quantity of old cement bags on the sea floor. They had been thrown out of some boat in the past.

Cement bags? That was it.

19

Along with its successes CEDAM International was having its problems. For example, the *Encyclopaedia Britannica Yearbook of Science and the Future, 1970,* contained an article on the work of CEDAM at the Sacred Well of Chichén-Itzá. The more people with an interest in diving and archaeology saw the organization's films, read about it in books, magazines, or newspapers, the more requests it received from individuals and organizations for assistance and participation in underwater projects.

And the problem was CEDAM International had no office facilities and no paid staff. For all intents and purposes Pres. Andy Rechnitzer was it. Lack of adequate financial resources was the barrier.

The organization now had six chapters, and Andy continued to urge them to develop an affiliation with an established museum to help with underwater archaeological projects. But other than the work of his own chapter on Catalina Island, response of the other chapters was sluggish.

To try to animate the chapters Andy furnished them with a list of films produced for the oceanographer of the navy and where they could be obtained. He listed books of interest and a description of a number of leading maritime museums. Aside from field trips and project work, he suggested the chapters hold semimonthly training meetings and provided a guide to subjects to be covered. They were: underwater exploration techniques; expedition planning and logistic support; excavation techniques; documentation methods and requirements; equipment operation and maintenance; first aid and diving safety; artifact preservation; photographic equipment and its use; literature search and report preparation; art and its uses; legal aspects; leadership; motion picture and slide series for lectures; lecturing techniques; advanced diving for underwater archaeologists; conference organization and management; underwater museology; maritime history; and information on cannons, anchors, coins, ceramics, bottles, and other indicators of time period and origin.

Finally he sought to inspire them by pointing out how in just a few years members of CEDAM International had made progress toward the organization's declared aims and goals as an indication of what more could be accomplished in the future. The list went like this:

Goal 1: Support museums established for the preservation, study, and display of artifacts from underwater.

Artifacts from four vessels, spanning a period from the sixteenth to the nineteenth centuries, recovered from Alacran Reef were on display in the CEDAM museums in Mexico. The National Institute of Anthropology and History Museum in Mexico City had Maya-Toltec treasure CEDAM had recovered from the Sacred Well of Chichén-Itzá.

Goal 2: Promote underwater archaeology and related ocean sciences as a rewarding hobby or avocation.

Adults and students participating in the expeditions had been indoctrinated in the methods and techniques required for properly exploring, locating, site survey, excavation, preservation, and documentation of archaeological items.

Goal 3: Stimulate increased public interest in controlled, systematic excavation of underwater historical discoveries.

Following the example established by CEDAM and its associates, several states had established positions for professional archaeologists and had taken positive action to assure controlled and proper removal and disposition of historical evidence underwater within the public domain.

Goal 4: Practice a diver's code of ethics that will encourage the conservation of underwater historical evidence and sites.

Sustained promotion by CEDAM International and similar organizations had helped in the increase in legal underwater excavations.

Goal 5: Conduct meaningful exploration and research, particularly in constructive support of underwater archaeology.

The findings of CEDAM International members derived from the sea and repositories of maritime history had brought evidence and knowledge back into the annals of readily available recorded history.

Goal 6: Engage in cooperative international exploration that will be of mutual scientific benefit to all participants.

CEDAM International had established a formal cooperative agreement with CEDAM of Mexico, its long-time affiliate. All expeditions had been a cooperative effort.

Goal 7: Disseminate to the scientific and lay communities through various communication media the results and analysis of field studies.

Through its documentary films and publications, CEDAM International had conveyed its findings to a broad range of European and Western Hemisphere scientific, technical, and lay organizations interested in this field.

Goal 8: Organize education, training, and field study opportunities for members and others interested in underwater archaeology and other marine sciences.

Strong emphasis had been given to the field training of members.

Andy said that underwater archaeology was a new and intriguing avenue of pursuit for many museums. They had experienced personnel in the methods associated with traditional terrestrial projects. In theory many local museums should be highly receptive to cooperative efforts. The number of CEDAM International members with experience in the field on expeditions was growing. Also, he thought, museums and others were coming more and more to appreciate CEDAM International's efforts to educate sports divers in becoming increasingly aware of the importance of respecting the historical value of underwater archaeological sites.

20

In the early 1970s CEDAM International and CEDAM of Mexico found themselves increasingly involved with planned or proposed expeditions.

If permits could be obtained, Robert Soehnlen and Paul Montgomery, of the Beloit Pipeline Company, responsible for pumping out the Chinkultic *cenote,* wanted to work with the two groups in an exploration of the Lake of the Sun in the volcano of Toluca. This was the highest lake in the world where diving exploration had been done, accomplished almost a decade earlier by CEDAM divers.

Dr. Nancy Farriss of the University of Pennsylvania, an underwater archaeologist, and Dr. Harold E. "Doc" Edgerton of MIT wanted to use sonar to explore the bottom of the lakes of Coba and Chunyaxche, looking for drowned cities and Mayan boats.

Prof. John Loret of Queen Mary College wanted to take core samples from the lakes of Yucatan to determine, through pollen analysis, the type of vegetation existing in the area for the last few thousand years and from this to determine what size population could have been supported.

Prof. Robert Newding of the department of biology, Texas A & M University, planned a two-year project to study the ecology of the Cove of Xelhá.

Bob Marx and Joe Kelly Hughes, now a CEDAM International vice-president, planned a special expedition for members to help them work certain wrecks in the Mexican Caribbean area.

Norman Scott was planning an excavation of a blockade runner at Cape Fear, North Carolina, with participation by members from the Washington, D.C., chapter of CEDAM International. He had also studied the history of Lake Guatavita, located near Bogotá, Colombia, at the Archives of the Indies in Spain and at the Public Records Office in London. From the latter he had obtained a report of Contractors, Ltd., a group that had excavated a part of the lake in 1903. Legend said the Chibcha Indians, as part of a ritual, years ago threw emeralds and gold into the lake. The lake, which Scott had visited, was situated at 11,000 feet. If Colombian government authorities would agree this would be an upcoming expedition.

Certainly one thing was becoming more apparent and that was the increasing difficulty to obtain permits for diving exploration. This was due in part to nationalism and in part to hostility of some archaeologists to the efforts of sports divers, regarded by them as amateurs.

In all this there was one promising new development in the United States which seemed to offer a possibility of beneficial expansion by underwater archaeological organizations affiliating with CEDAM International in furtherance of mutual goals.

A small group of amateurs and scholars interested in the field in the State of New York in 1970 had formed the Underwater Archaeology Association (UAA) with goals similar to those of CEDAM International. Two years later, though retaining their individual identity, they affiliated with CEDAM International, their by-laws stating that all members were also expected to belong to CEDAM International. This group was recognized as a new chapter.

UAA established a cooperative program with Elmira College, diving on sunken canal boats in the Finger Lakes and exploring Glass Factory Bay, former location of a nineteenth-century glass factory, establishing a conservation laboratory and developing an undergraduate program in nautical archaeology at the college.

Meantime, each summer the usual joint expeditions were held.

During this period all explorations were on Chinchorro Reef. Results were not spectacular, but no matter. Members could be inspired by the work of others like that of pioneer Bob Marx who in CEDAM International's bulletin described how earlier he had found the wreck of two of Columbus's caravels, lost on his fourth and final voyage to the New World.

Marx, an avid student of old shipwrecks and their history, had read Adm. Samuel Eliot Morison's Pulitzer Prize winning book, *Admiral of the Ocean Sea.* In 1940 Morison had led an expedition to St. Ann's Bay, Jamaica, and his book contained a chart of this bay showing the estimated location of the wrecks.

Off Jamaica lay the sunken city of Port Royal in an area scheduled to be dredged for construction of a deep water port. Marx had gone to Jamaica to work on excavating this site, fighting a deadline against the dredging. While there he met Charles Cotter, a plantation owner who had helped Admiral Morison on his expedition and ever since had tried, without success, to find someone to dive on the estimated location of Columbus's ships. Marx was unable to interest his employer, the Jamaican government, in this venture, especially because there was little likelihood of finding treasure. He told Cotter he thought he could spare a few Sundays to look into the matter.

This is where CEDAM International's eager divers could pick up their ears, and maybe their gear, to enter on challenging projects of their own. To conduct the search there was just Marx, his wife, and two diving friends.

Using long metal rods as probes, excavating by hand and with buckets, after six hours they finally hit an object which turned out to be a wooden beam. Bob was convinced it was piling from an old wharf, but his wife had a hunch. Going back and digging some more, he felt wooden pegs in the beam and knew then and there it was part of an old ship.

On the basis of this evidence, he enticed "Doc" Edgerton down from MIT to use a sonar device he had designed for locating archaeological material under mud. Within an hour on the site, two positive sonar contacts had been made.

On the advice of famous nautical archaeologist Dr. George

Bass, an honorary CEDAM International advisor, Marx next contacted Dr. John Saunders of Columbia University who had invented a coring device. This resulted in Saunders' assistant, Bob Judd, coming down to Jamaica with a corer to work for a week.

It took all the strength of the two men to use fifty-pound hammers to drive the coring tube into the sediment; visibility in the water was almost nil because of the sediment stirred up. Samples of material from the corer were placed in water-filled jars and tagged for location and stratigraphical depth. Holes made by the corer were covered up to prevent invasion by sea teredos (ship worms).

Datable material was sent to experts in England, Spain, and the United States for examination. Marx said the final test was submission of the findings to Admiral Morison and Sr. Mauricio Obregón, a Colombian expert on Columbus. He said both agreed they had found wrecks of the discoverer of the New World.

From stories like this, amateur underwater archaeologists could dream of forming their own expeditions. But, of course, treasure remained at the back of many minds, poisoning the well, building up resistance on the part of the professionals.

21

For its expeditions in the early years, CEDAM rented Capt. Argimiro Argüelles's boat and even named a shipwreck after him. Argüelles owned the copra plantation that fringed the shore in front of the jungle at Akumal where CEDAM often established its forward base, slinging their hammocks and pitching their tents forty miles away from nearest civilization, Cozumel Island, across the water. It was a virgin area.

In 1960 Captain Argüelles told Pablo Bush he was getting too old to work the copra plantation any longer. He said Akumal was the best cove and location on the Mexican Caribbean and offered to sell it to CEDAM's president. A shrewd businessman as well as an explorer and conservationist, Pablo bought it on the spot. Not only this, but over the next years he bought a good deal more of the coastline both to the north and south of Akumal looking ahead to tourist possibilities.

By 1967 a land company called Promotora Akumal Caribe had been formed and development started. It was not easy to get workmen or other personnel at this isolated spot where there was no electricity, little water, poor transportation, and a host of other problems. Little by little these difficulties were overcome.

In the cover story for the August 1973 issue of *Skin Diver,* its writer and CEDAM International member, Bill Barada, entitled his piece: "Akumal—A Diver's Dream Come True." He had been to Akumal almost ten years earlier on a CEDAM expedition and had dived in "the gin-clear water beyond the reef where fish were so tame and so abundant we sometimes had to push them away before we could take their picture." At the time he had written an in-depth lead article about his experiences there entitled "CEDAM International" for *Oceans.*

Now, standing on the same spot by Akumal's lagoon, Barada described this "dream come true." He said, "The scene was still of a virgin, primitive area. Palm-thatched huts of a Maya-type village nestled among the coconut palms. Nearby, a building of native rock construction and a thatched roof disguised a gorgeous modern dining room . . . Above, on a rocky bluff overlooking the entire lagoon, were two-story, Maya-type bungalows which served members of the newly created Club De Yates Akumal Caribe."

Pablo had taken careful steps to protect the environment, important in itself and as an example to other developers who were sure to follow. The beach, with its sugar white sand, had been kept clear. It was still a place dominated by the blue Caribbean, the tropical sun, the tradewinds, the coral reefs, the crash of waves, graceful palms, iguanas sunning on the rocks, and the jungle.

There was one problem. When Pablo had proposed the tourist development of Akumal to candidate Echeverria, soon to be president of Mexico, he was told this was too much land to be held by one person or even a group of persons. After two and a half years of bickering, they were forced to return 2,000 hectares to the government, keeping 60 hectares for the development. One good thing was that Pablo managed to have the cove of Xelhá, part of this land, converted into a national park. The arguments advanced in all this played a major part in influencing the Mexican government to develop Cancún, north up the coast, from a sleepy small town into an east coast tourist competitor to Acapulco on the west coast. More and more hotels were being built on Cozumel Island.

What mainly had opened up Quintana Roo to the outside world was a new, flat two-lane, hard-surface road, straight as an arrow, cut through the jungle. From Cancún it ran, just in from the coast, to Mexico's border with British Honduras, later to become the tiny independent nation of Belize. Tourists could now circle the Yucatan peninsula, going through the heart of the ancient Maya sites, Chichén-Itzá, Uxmal, Dzibilchaltum, Coba, and Kohunlich.

In the years ahead increasing tour groups from places like Japan and Germany would fly into Cancún, where buses would take them on a side trip to the walled city of Tulum, then stopping briefly at the Cove of Xelhá and having lunch at Akumal as part of the one-day excursion. Divers and fishermen could strike for the coast on new small feeder roads, reaching spots on the water previously accessible only by boat.

But in the early 1970s Akumal was still remote. Anyone with a light plane could land on the rocky strip at Tulum where, if they first circled low over Akumal, a Jeep would be sent for them. It was possible to get a commercial plane to Cozumel and take a ferry over to Akumal. Or one could fly to Merida and drive down in a rented car, but there were few gas stations, little drinking water, and plenty of potholes with virtually no automobile repair facilities.

At the new Hotel Club Akumal there were no telephones, TV, newspapers, or radio entertainment. Trade winds furnished the air conditioning except for a few window units. The electricity, generated on the grounds, was erratic. Overhead fans and mosquito nets kept these pests at bay. A radio wireless, used twice daily, connected Akumal with the outside world.

CEDAM museums in both Mexico City and Progreso had been closed and the cannon, coin, utensils, dishes, jewelry, muskets, anchors, and all the other artifacts brought to Akumal where a new CEDAM museum was to be opened.

One innovative idea was the development of an underwater museum, literally underwater, with an anchor and some old cannon placed on the reefs in shallow water in Akumal's lagoon. Items such as these normally need a great deal of preservation effort to keep them from crumbling in air when brought up after years in the water. One thought was it would be unnecessary to go through this process with the offshore museum objects. Also in this simulated graveyard of the sea skin and scuba divers, or even visitors using the hotel's glass-bottomed boat, could get an appreciation of how these objects looked when first discovered.

CEDAM International's annual 1971 lecture program was held in Long Beach, California, to coincide with the meeting of Oceans '71. Those present viewed a new film made by Gordon Inman, in charge of the Purex water treatment system on the 1967-1968 Chichén-Itzá expedition, entitled *Treasures of Yucatan.* The film dealt principally with the archaeological work of the expedition and touched on the technical aspects of the water treatment as a means of clearing water for diver's visibility. One other development at this meeting was a request by Joe Kelly Hughes for members to forward to him any research they might do on ship sites so that CEDAM International could begin to have its own archives.

The 1972 meeting, hosted by the local chapter, was held in Washington, D.C., at the Museum of History and Technology of the Smithsonian Institution. A number of artifacts from the Matanceros wreck were presented to the museum.

It was inevitable, with development of the facilities at Akumal, that the first meeting outside the United States should be held there in 1973. The meeting was a huge success. Members could swim, dive, fish, visit the ruins, and examine the artifacts in CEDAM's own museum.

Films were shown every night, and during part of each day there were talks on underwater explorations from the Sinai peninsula and the Red Sea, the Alto Plano of Bolivia, wrecks in the Bahamas and Lebanon, to reports on CEDAM International's own expeditions—the *Tweed,* Chinkultic, and the like.

But all was not underwater material. One of the most fascinating talks was given by anthropologist Gertrude "Trudy" Blom, known as "the Mother of the Lacandonian Indians." She spoke of the past, present, and future of this vestige of the original Maya population, once estimated to have been more than five million, now reduced to a few hundred.

These Indians, living much as their preconquest ancestors did, were settled in small clearings in the rain forest surrounding Lake Naja in the State of Chiapas. They traveled about the lake in dugout canoes and lived on a slash-and-burn subsistence economy in the jungle. Blom told members these Indians needed help now that their future might be assured. In addition their past should be preserved.

CEDAM International's executive committee agreed that the organization wholeheartedly supported Blom's work. They set up a fund in her name in order to accept tax-free donations from those wanting to support her activities and the maintenance of the valuable library under her direction at San Cristobal de las Casas in Chiapas.

And as to Akumal, it was agreed henceforth it would be regarded as the "world headquarters" of both CEDAM International and CEDAM of Mexico.

22

By the early 1970s it was becoming increasingly apparent that new problems were emerging for sports divers like CEDAM International members interested in archaeology as well as for terrestrial archaeologists interested in underwater sites.

One facet of this was contained in a paper by two archaeologists, Neil F. Marshall and James R. Moriarity, reproduced in the Spring 1971 issue of the CEDAM International bulletin. They said in 1950 swimmers using aqualungs had discovered Indian artifacts off the shore of the Beach and Tennis Club at La Jolla, California, and brought them to their attention.

The discovery of artifacts in offshore regions, they said, immediately added to the complexity of the problem of meshing underwater data with dry-land data. They said, "Further study of underwater sites will require an increasing application of the tools of physical oceanography."

In this the writers were obviously not thinking about sports divers playing any significant role other than perhaps the incidental discovery of artifacts.

Though a gulf was growing between sports divers and archaeologists, some thought ways could be found to bring them together. This was exactly what CEDAM International had been preaching. Samuel P. Townsend, assistant director for the Division of Historic Sites and Museums for the State of North Carolina's Department of Art, Culture, and History, laid down guidelines in a paper he presented in January 1973 at the Fourth International Conference on Underwater Archaeology held at St. Paul, Minnesota.

Each side, he said, accused the other of being greedy. There was the need to define common goals in order to coexist peacefully. His paper was a case study of how this had been accomplished between the State of North Carolina and a nonprofessional group, Underwater Archaeological Association, Inc., a diving organization. The absence of "a treasure hunting motive" had been "probably the most important factor" making for salvor/archaeologist cooperation.

There had to be an understanding, he said, of what both sides *want* and *fear*. "The archaeologist *wants* first to study, scientifically, past human remains and activities to acquire new knowledge; artifact collecting, as such, is secondary to him. The archaeologist *fears* that the salvor will damage or destroy sites and steal artifacts. Most salvors *want* to recover artifacts for profit or for the fun of collecting and prefer that their activities be adventurous and interesting. They *fear* that archaeologists or government authorities will hinder or stop them from working sites or, at least, keep all the best of the artifacts and other benefits."

A fundamental for coexistence in North Carolina, he said, was the recognition that both sides needed each other. The salvor organization needed a permit to work state-owned sites, and the state needed volunteer assistance because of inadequate funding for archaeological underwater work.

Results had been fine. Salvors had "internalized the values of the archaeologists" and meaningful archaeological work had been done at little cost to the state.

The most important single thing the organization had done was to spread the word among sport hobbyist divers that archaeological sites are important and should be approached scientifically.

In another paper, Claude Hull, of Underwater Archaeological Associates, Inc., described the role of research of the colonial period as related to North Carolina underwater archaeology. These principles could be applied anywhere in this relatively new field.

He said researching colonial shipwrecks could best be compared to reconstructing a gigantic puzzle. A search of the colonial records of the state and nearby states, correspondence, and old newspapers of the period all could be helpful.

The Public Record Office in London was a rich source of information, including ship's plans. The Archives of the Indies in Seville, Spain, had lists of vessels entering and leaving Cadiz during their colonial period with information on ports, cargoes, and the "owner's mark." Colonial maps and charts in the U.S. often contained data on shipwrecks.

CEDAM International had become very familiar with these identification problems. Andy Rechnitzer had pointed out that cannon were helpful, but they were often traded from one ship to another; coins gave dates and countries, but these passed from hand to hand. Bottles were good when they had the firm and the date visible, but probably best were hallmarks on silver plate. From these the exact date and place of manufacture could often be determined.

For the sports diver and the underwater archaeologist other problems were coming to the forefront in the U.S. Broad aspects were discussed at the Fourth International Conference on Underwater Archaeology by George R. Fischer, research archaeologist at the Southeast Archaeological Center, Tallahassee, Florida.

A major part of the problem was its great scope, he said. The National Park Service administered substantial areas, especially in the National Recreation Areas, of submerged lands rich in underwater archaeology. In just four of these areas, Gulf Islands National Seashore, Biscayne National Monument, Fort Jefferson National Monument, in the Dry Tortugas, and the Everglades National Park, there were a total of some 600,000 acres underwater. He said a recent study had revealed evidence of nearly 200 shipwrecks in the 50,000 acres of water around Fort Jefferson alone. Obviously the service was responsible for an incredible number of historic shipwrecks.

Conducting "a thorough and accurate survey" was a "staggering proposition." He said the National Park Service's primary concern was protection from "unauthorized exploitation and destruction, either by sports divers or illicit organized salvage activities." Park rangers needed information to protect these sites.

He also pointed out that the federal act for the Preservation of Antiquities of June 8, 1906, was a congressional mandate that antiquities on federally owned or controlled land was the property of the American people. Also it was the law that excavations could only be made "by qualified archaeologists operating under strict controls and adhering to recognized professional standards."

On another front the period was seeing a new thrust toward the sea on the part of those searching for oil. One aspect of this was regarded as beneficial to the expanding field of professional underwater archaeology though little to do with the sports diver. The government now would require, prior to drilling, an archaeological study to be made of what was termed a "cultural resource." If such was found, underwater archaeologists would go to work. This was taken by many as an indication of new recognition by the U.S. government of the significance of underwater archaeological sites as a national legacy.

Different states were paying new attention to underwater objects. Florida was one state that had adopted strict antiquity laws. Writing of this in *Oceans,* in his article on CEDAM International, Bill Barada said, "Florida's laws are so rigid and encompassing that it is illegal for a diver to even search for sunken objects without a $600-per-year search contract with the state, and he must have a $1,200 permit before he may recover an object. This law applies to old bottles, bones, arrowheads, etc. . . . a law so all-encompassing it is not only unenforceable, it is self-defeating. Its effect has been to drive souvenir hunters underground and create a black market in artifacts."

Barada thought CEDAM's program to invite sports divers to participate in archaeological expeditions under the direction of professionals was much more realistic than Florida's "hands-off" policy.

Some thought CEDAM's approach should be expressed in law. Howard Shore wrote a detailed technical paper entitled

"Marine Archaeology and International Law: Background and Some Suggestions," which appeared in the May 1972 issue of the *San Diego Law Review*.

He said, "Professional archaeologists are coming to realize that dedicated amateurs, working under expert supervision, can make an important contribution to the acquisition and protection of information concerning past civilizations."

He went over CEDAM's formation, followed by that of CEDAM International, the organizational goals and achievements. Shore noted the increasing frustration of such organizations when confronted with restrictive coastal state laws for access to underwater archaeological sites.

He suggested that an international convention should adopt a principle declaring that certain organizations "be given international recognition for their expertise and sincerity in achieving the objectives of marine archaeology." They should be given special consideration "in the granting of deeds of concession for exploration and excavation conducted within coastal state jurisdictions."

For an organization like CEDAM International, he said, permission to conduct marine archaeological research should be "nearly automatic."

Entrance to the Hotel-Club Akumal, Quintana Roo, Mexico. CEDAM's Underwater Museum is located in the building on the left. Just beyond are the clear, blue waters of "Turtle Bay," forward headquarters for CEDAM's early expedition. (Photo by Dave Friedman)

CEDAM divers explore Palancar Reef off Cozumel Island at ninety feet. (Photo by Julian Lindenauer)

CEDAM International president "Andy" Rechnitzer *(right)* points to markings on old cannon brought up from Alacran Reef. A workman and Joe Kelly Hughes, a veteran CEDAM diver *(left)*, look on. Rechnitzer was scientific advisor for the submersible that made the deepest dive ever into the sea. (Photo by Pablo Bush)

CEDAM founder Don Pablo Bush Romero surfaces from a dive into the Sacred Well of Chichén-Itzá with the skull of a young boy, a Mayan sacrifice. (Photo by Genaro Hurtado)

Pottery, bones, and various artifacts are lined up beside the Sacred Well of Chichén-Itzá by archaeologists seeking clues to the past from the watery capsule of history. (Photo by Genaro Hurtado)

The reef off Quintana Roo, Mexico, second longest in the world, holds many wrecks pushed there by the force of hurricanes to break up under the battering waves. Other wrecks, new and old, litter the bottom. (Photo by Genaro Hurtado)

The small Mayan building on the promontory in the center of the picture, part of the ruins of Tulum, may have played a role in both coastal navigation and defense for this ancient race in days long gone. (Photo by Lorane A. Wilson)

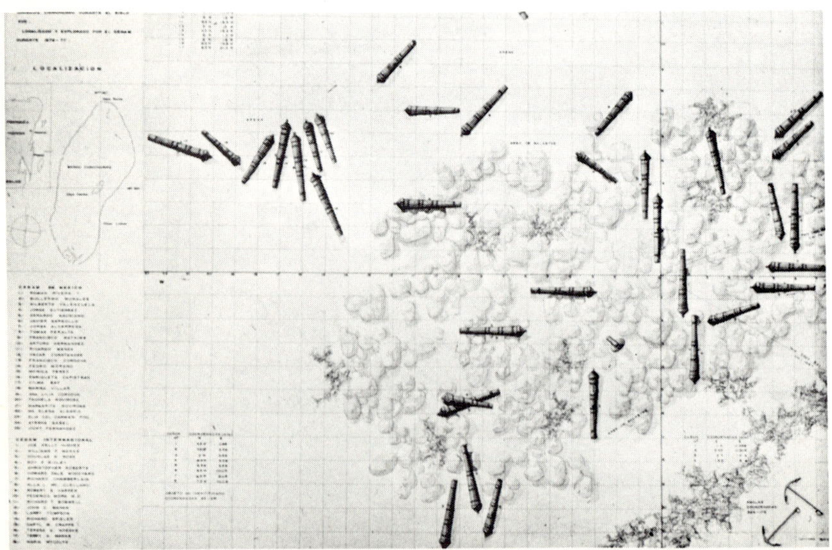

Mexican architect Vilma Bay, one of CEDAM of Mexico's female divers, made this drawing of cannon scattered over a wreck site on Chinchorro Reef, aptly called the Forty Cannon Wreck. Names of members of the 1978 expedition are listed on the left. (Photo by Julian Lindenauer)

This is artist Vilma Bay's conception of how the mystery ship known as the Forty Cannon Wreck looked before piling up on Chinchorro Reef long ago. (Photo by Julian Lindenauer)

Dr. Richard Spieler and Dr. Teresa Noeske, of the Milwaukee Public Museum, examine a fish specimen they collected while on a CEDAM International expedition to Chinchorro Reef. They were also studying the ecology of the Quintana Roo coast. (Photo by Bob Harper)

A small statue of the Virgin of Guadalupe, CEDAM's protector, looks over the bay at Akumal. In the late 1950s a large sculpture of the Virgin was placed under the sea at Acapulco as the New World's first underwater shrine. (Photo by Dave Friedman)

Spanish cannon recovered from the sea stand guard at Akumal, Quintana Roo, Mexico. Other old cannon and anchors are in the nearby, crystal-clear offshore waters as part of CEDAM's underwater museum. (Photo by Lorane A. Wilson)

Author Earl J. Wilson stands beside an enormous old Spanish anchor recovered from the sea by CEDAM divers and now part of its museum at Akumal. (Photo by Lorane A. Wilson)

The sun strikes the Kukulcan Pyramid at Chicheń-Itzá at dusk on the equinox, the sunlight—as planned—forming for a few moments a weaving pattern down the steps on the left, coming to rest on the enormous head of a stone serpent at the base. From here Mayan maidens were led to be sacrificed in the nearby Sacred Well, target of several CEDAM expeditions. (Photo by Luis E. Arochi)

Three small fish in the muzzle of the cannon on the right form a human face. Diver is examining site of the Forty Cannon Wreck. (Photo by Julian Lindenauer)

Posed in Akumal for this 1978 photo are *(left to right):* Bernarr Macfadden, Dallas, Texas, president of CEDAM International; Joe Kelly Hughes, CEDAM director of operations; Don Pablo Bush Romero, El Paso, Texas, founder of CEDAM; Michael J. Kelly, Woodstock, Illinois, CEDAM board chairman; Javier Gargollo, of CEDAM of Mexico; and Chuck Ennis, San Antonio, Texas, one of the founders of CEDAM International. (Photo by Lorane A. Wilson)

23

In the early 1970s Andy Rechnitzer was fond of telling members there was a good chance in the decade ahead of their finding the first sunken city in the Western Hemisphere.

He pointed out geologists had established that 8,500 years ago the sea was 100 feet below its present level. Indian midden sites had been found along the southern California coast as far as a mile from shore. Using carbon-14 dating and oxygen-18 temperature determination, it was possible to learn when these sites were occupied and what the temperature of the ocean was at that time.

As to the Yucatan, he said he believed the Mayas were in the region 7,000 years ago even though artifacts dated them only 3,600 years back. The Mayas were so advanced and adept at building stone structures it was reasonable to assume that some of their cities existed on the shores of the New World when the ocean was lower and that divers might find evidence of this in 100 feet of water.

And what about Maya boats? Elizabeth P. Benson, who had visited Akumal while doing research for her authoritative book, *The Maya World,* wrote: "There must have been a great deal of trading by sea, from Tabasco around the Yucatan peninsula to the Gulf of Honduras. There are sixteenth-century Spanish accounts of encountering large dugout canoes with sails making their way around the coast, laden with a variety of goods." Here would be something exciting for CEDAM International divers to look for! Wouldn't it be something if they could be the first to find one of these old Mayan trading boats under the sea!"

There were mysteries. Andy pointed out that the little Mayan buildings along the coast were usually oriented so that the doorway faced an opening in the reef. At Tulum there was one building with two windows. He said if you lined them up you were right on the channel through the reef, an aid to seafarers at night, or even in the daytime when wind conditions could make it difficult to see the clear channel. He said, "We know that Cozumel was occupied. They had to run back and forth so there must be some evidence along the coast."

Andy suggested some of these possibilities to underwater archaeologist Dr. Nancy Farriss and "Doc" Edgerton, using CEDAM International divers to team up in the search.

A CEDAM International member with similar interests was Talbot S. Lindstrom, an international lawyer with an office across the street from the White House in Washington, D.C. In 1971—with a fellow lawyer, Steve Proctor, and a marine archaeologist and geologist, John Gifford—he had formed SEAS, the Scientific Exploration and Archaeological Society.

These three men had been fascinated by the similarities in architecture, pottery designs, legends, gods, inscriptions carved on rocks, and other data suggesting the possibility that centuries before the discovery of America by Columbus, Old World colonies existed in the Western Hemisphere.

They thought through nautical archaeology it might be possible to locate port sites with artifacts to prove cultural and trade links across the Atlantic in those ancient days. Also, with rampant development they were afraid many potential port sites would be destroyed before they could be carefully explored and evaluated.

SEAS was founded basically as a survey organization seeking interesting leads that would warrant detailed follow-up study by marine archaeologists and other professionals. Lindstrom, who speaks nine languages, spent most of his spare time with his wife, also a qualified scuba diver, underwater in the Western Hemisphere or abroad looking for clues to the past.

This interest led him to Akumal, where he and his colleague,

Steve Proctor, photographed much of the shoreline in the vicinity from the air as an aid to follow-up underwater study. From this they identified three interesting sites: Xelhá, a possible harbor; Tancah, with its reef-protected bay; and the Chunyaxche complex of inland lagoons and waterways.

In the summer of 1974 both CEDAMs and SEAS worked in support of Dr. Farriss and "Doc" Edgerton to look for evidence of Mayan maritime commerce and coastal wreck sites. Expeditionaries were assigned in teams to search likely spots along the Quintana Roo coast.

Andy, the Fifields, Alfonso Romero, and some others went to explore the lakes of Coba. This huge ruined Mayan city, a ceremonial center, was nearly swallowed by the jungle. Vestiges could still be seen of many *sacbe*s (highways) radiating from this city to other parts of the Mayan empire. Some were forty feet wide. The main pyramid was taller than that of Chichén-Itzá.

No one could figure out where the Mayas obtained the great blocks of limestone used in building the city of Tulum, thirty-nine kilometers away. The explorers thought perhaps the lakes at Coba were old quarries used for this purpose but this did not check out.

On the average the lakes were seven meters deep, the bottom covered with so much undergrowth as to prevent use of sub-bottom profilers in the search for old Mayan boats. The group waded around the lakes making measurements and using plumb lines to record depths. They did find remnants of an old boat dock. The water was too fetid for diving.

In the fall of 1974 Mary Mykolyk, an architect and CEDAM International member living on Cozumel, had gone to the windward side of the island with her scuba diving friend, Ronald Oberst, to get away from the tourists at an isolated spot called Chen Rio, Mouth of the River.

There they donned snorkel masks and went in to check the sea conditions. Not far from shore Ron saw something unusual on the bottom, a smooth green rock. As he dove for it he later said, "The thought occurred to me that there are no native rocks on

Cozumel other than the limestone formation which to my understanding composes the strata of the entire Yucatan peninsula as well as this island." He said as soon as he gripped the stone he knew it had been handworked. Before he reached the surface he knew it was jade. Mary realized at once the archaeological importance of the find. They went back to the beach for their scuba gear.

They decided they would only pick up the scattered pieces lying on the bottom, but not the worked stones embedded in the coral rock. They felt the former might be lost in the shifting sands, while the latter would be clues for future archaeological research.

When the find was reported to the Mexican authorities, the director general of the National Institute of History and Anthropology sent the director for the Yucatan peninsula, Don Norberto Gonzalez, a terrestrial archaeologist, to look at the find underwater. For this task Don Norberto rapidly learned scuba.

The collection included more than sixty stones of different coloration ranging from very light yellowish green to dark blue-green and black. They varied in size from an eight-inch long axe-shaped artifact to a one-eighth inch diameter bead. Axe-head and bead shapes dominated the collection. The most distinctive piece was a small blue-green bead carved to depict the human face.

It was a significant find. These worked stones were presumed to have dated back to a pre-Colombian culture and were the first evidence recovered from the sea to support the theory of the sea-trading Maya. They were apparently lost from a boat crossing through the surf.

A follow-up expedition to the 1974 survey was arranged for 1975 with plans for Dr. Farriss and "Doc" Edgerton to narrow their search, using acoustic sensors, to the three candidate areas along the coast where Mayan canoes might have foundered. This expedition was supported by SEAS, *National Geographic,* Alcoa Foundation, and the Desco Corporation with CEDAM International divers again aiding in the underwater work.

There were a variety of problems complicating the task.

First, they had no old Mayan documents to guide them to shipwreck locations, not even modern fishermen's tales. They didn't really know what kind of boats or canoes they were looking for. Since these craft had to be of shallow draft to get through and around the reefs, they would not carry ballast, such ballast stones on the bottom often alerting divers to wreck sites of ships of the colonial period. Mayan boats had no cannon, no large anchors, and few metal objects. Wood in tropical waters soon disappeared under onslaughts of the teredo. What were they looking for?

On the other hand, there was increasing evidence to suggest that by the time of the arrival of the Spanish the east coast of the Yucatan peninsula had a flourishing long-distance maritime trade system. If so, there had to be something under the sea. The problem was to find it. The 1974 survey expedition had settled on three likely spots along the coast and these were again the target.

First was Chunyaxche, explored by Pablo Bush, Emmett Gowen, and other CEDAM pioneers seventeen years earlier. They had noted that some of the structures along the lake were accessible only by water. The survey of the lake bottom the previous year had revealed soft sediment, ideal for capturing and holding a wooden vessel against the ravages of time and the water. Unfortunately, they found the sediment had poor acoustical properties, making the sonar useless.

Tancah had a hard bottom, and the search here was narrowed to the opening in the barrier reef where the surf could have caused accidents to small boats in the past. They found a narrow slit underwater, dove through it, and emerged in an enormous cave where they found a few potsherds. Most successful was Xelhá inlet, thought to be the best natural harbor on the Quintana Roo coast. Test holes were excavated and archaeological material was recovered.

All this was described in an article entitled "Maritime Culture Contact of the Maya: Underwater Surveys and Test Excavations in Quintana Roo, Mexico," by Dr. Nancy M. Farriss and Dr. Arthur G. Miller, Center for Pre-Colombian Studies,

Dunbarton Oaks, Washington, D.C., in the May 1977 issue of the *International Journal of Nautical Archaeology and Underwater Explorations,* published in London.

The article said material recovered was ceramic, several fishing sinkers, and well-preserved organic material. They found a charred timber projecting into one test hole, possibly part of a shore structure. Excavations were limited both by time and the Mexican government permits. On the basis of what was found they thought future systematic excavation of Xelhá inlet would be rewarding.

"The type of archaeological remains we believe most prevalent at Xelhá is 'port refuse'—the material lost and discarded by mariners that tends to accumulate in any busy harbor. . . . Such material would provide a valuable record of Mayan maritime activities, including seaborn contact with neighboring areas. The grotto shrine, which unfortunately by now has been thoroughly looted, indicates that the area had religious significance and may have been one of the east coast centers of pilgrimage. The peninsula's defensive wall suggests that the area may also have been the target of less peaceful visitors, perhaps seagoing raiders like those depicted in mural paintings in the Temple of the Warriors at Chichén-Itzá."

Perhaps, but this exploration remains to be accomplished. The Mayan seafarers continue as elusive as ever. The tireless ocean covers all.

24

In the mid-1970s CEDAM International members were intrigued to read of the involvement of several of their fellow members, including their president, in confirming the identification of the Union ironclad ship, U.S.S. *Monitor,* sunk off Cape Hatteras in the Civil War.

The search began in August 1973 when an expedition, put together by John Newton, marine superintendent of Duke University's Marine Laboratory, and Gordon Watts, of the North Carolina Department of Cultural Resources, set out to map the ocean bottom in a 5 by 14-mile rectangle stretching southeast from the *Monitor*'s last reported position. With them was CEDAM International advisor "Doc" Edgerton, "father of the underwater camera."

A wreck, possibly the *Monitor,* was located at the end of the two-week expedition, leaving only three days for underwater photographs. A six-month study convinced them they, indeed, had found the famous ship. This was based on painstaking measurements of their photos against the original architectural drawings of the *Monitor*.

Watts, only twenty-eight, told skeptical newsmen, "The problem is that everybody thinks they know what a shipwreck looks like. They've seen them in the movies with the sails still flapping underwater and all that. And what we have here instead looks to be an obvious pile of trash."

The following year, during the first week of April 1974, search operations again were conducted, this time by Alcoa's research vessel, *Seaprobe*. It was on contract to the navy. They wanted to evaluate its shallow water search capabilities. CEDAM International members included the officer-in-charge, Comm. Colin Jones, USN, Pres. Andy Rechnitzer, and, again, "Doc" Edgerton.

Everybody knew about the *Monitor*. The Confederate iron-clad *Virginia,* later called the *Merrimac,* was playing havoc with the wooden Union fleet. The *Monitor,* the first ironclad built with a movable gun turret, was built in 101 days, designed specifically to stop the *Merrimac*. Her design symbolized the end of the wood warship.

Launched in January 1862, the two ironclads were hammering at each other at point-blank range within six weeks at Hampton Roads, Virginia. After four hours the *Merrimac,* short of ammunition and leaking, withdrew into Norfolk where she was later burned and scuttled by Union forces.

On December 30, 1862, the *Monitor* was being towed south to join the fighting in North Carolina. On that date, off Cape Hatteras, "graveyard of ships," a violent storm arose. The *Monitor* went down with a loss of sixteen of her crew. For 100 years the location of the wreck remained a mystery.

Seaprobe made numerous systematic passes over the hulk recording the scene with both television video and still cameras. More than 1,200 still photographs for a mosaic were taken of the "cheese box on a raft" to establish positive identification. The ship was upside down in 220 feet of water about 16 miles off Cape Hatteras.

A few months later the secretary of commerce, with the approval of the president, designated the resting place of the *Monitor* a marine sanctuary to protect the wreck from treasure hunters or salvagers. And there she remains.

Few people have seen her. One of those who has was CEDAM International advisor Larry L. Booda, editor of *Sea Technology*. Certain events since he had passed the age of fifty, he wrote, were "seared in memory." One of these times was when he piloted an F-4 Phantom II fighter at Mach 2.05. A second was witnessing the parade of the tall ships in New York Harbor on July 4, 1976.

The third was his underwater visit to the wreck of the *Monitor*. In the summer of 1979, while archaeological investigations were being made, he went down in the Harbor Branch Foundation's Johnson submersible, *Sea-Link II,* designed and made by Ed Link, another CEDAM International advisor. Booda spent one hour and twenty minutes circling the wreck at 210 feet, a scene he says he will never forget.

In the summer of 1975 Lt. Comm. Joe Kelly Hughes, who had been serving as naval attaché, La Paz, Bolivia, left the navy to live at Akumal with the twin assignment of running the club's dive shop and representing CEDAM International. Born in Mexico, he spoke fluent Spanish, had served abroad with the navy as a diver, and while in Bolivia had explored underwater in Lake Titicaca and the Amazon. Hughes had been active in many past CEDAM expeditions and was now a vice-president. In his new position he helped organize the June 1975 expedition which had an objective of scouting the Quintana Roo coast in support of the Farriss search for Mayan ports and boats, as well as looking for coastal shipwrecks.

The thirty-member group included Andy Rechnitzer, Pablo Bush and his two daughters, Myrna and Laura, board member Taylor Caffrey, from New Orleans, and Tom and Marilyn Fifield, of Milwaukee, long-time CEDAM explorers, who flew down each year in their own small plane, landing at Tulum.

First target was a shipwreck a local lobster fisherman, who went along as a guide, had told them about. They found it, just as he said, and named the new site *Tres Anclas,* or Three Anchors.

Basic reconnaissance was carried out from the beach toward the reef, using range poles as reference markers. Lines were used to determine the distance between the artifacts after they had been marked with buoys. Taylor Caffrey was one who swam about, fighting the surf and holding a hand compass to determine relative magnetic bearings. An underwater metal detector turned up various "hot" areas beneath the coral. Excavation was left for another day.

From the evidence Hughes and the others tried to visualize

how this ship, which they judged to be a square-rigger of several hundred tons, had come to grief. The anchors formed a straight line on a bearing of 270°. From the position of these anchors could be seen the tremendous force of the ship as it drove over the reef. The outer anchor, poised with its fluke embedded in the coral, seemed almost poetic to Hughes. Later he wrote: "If ever violent force and imminent destruction is to be captured in a monument, this anchor is that—giving mute testimony of the final, frantic struggle of the ship's crew while destruction was upon them."

They thought the ship, caught by a southwesterly gale, had been driven on the reef as the crew, in a last frantic effort, dropped the first outer anchor before the hull was smashed on the coral, opening it to the sea while huge waves pounded above, dropping tons of seawater on the deck. Then somehow the ship was swept over the reef as they dropped the two additional anchors to keep the ship lined up as it ran aground in the shallows on the inner side of the reef.

Afterward, probably in an effort to lighten and refloat the ship, the six cannon they found were dumped over the side.

One day when the wind blew up, making diving on the *Tres Anclas* difficult, the expeditionaries shifted their attention inland almost directly behind the beach from the wreck. They carried their diving gear and a rubber boat 100 yards through the jungle to a large limestone sinkhole called the Hidden *Cenote.* One member got in the rubber boat while the others circled the oblong lagoon, going in at intervals and passing their data up to the man in the boat. A large underwater cave with an entrance 60 feet wide and 15 feet high was discovered. They entered for 30 feet, as far as they could see with natural light. No artifacts were found but the place was noted for future exploration with proper underwater lights.

Half of the group took off to investigate the Lake of Chunyaxche. Though they found no artifacts, they did their bit as conservationists. Construction of the Boca Paila truss bridge had opened up the area so the inhabitants no longer used the canals. As a result of this and weather factors, a sandbar had built up across the exit leading into the lake. This, in turn, prevented the passage of many saltwater fish from entering the fresh water of the inner lake as had been their custom. Many of these fish in a weakened condition were observed swimming near the surface. The divers armed themselves with shovels and attacked the sandbar, finally clearing it so that the water began to flow.

Pushing their inflatable boats along a shallow channel, they crossed the lake, stopped at a small temple on the other side, and began an exploration through a meandering channel thick with mangrove. To lighten the load on the rubber boats many of the team went over the side and swam using snorkels for the six miles of their exploration. No artifacts were located, though in the fresh water they did see numerous saltwater fish.

This phenomenon was especially apparent when they explored Xelhá where saltwater fish swam side by side with freshwater fish. Freshwater penetrating the porous cap of limestone formed underground rivers that deposited great quantities of fresh water into the sea producing this unique ecosystem. At times when the sun was right the thermocline separating the two layers of water shone like a mirror. Diving through this interface set up a turbulence that, though both the salt water and freshwater were crystal clear, for the diver momentarily made everything go out of focus.

In the summer of 1975 another expedition was at work on Chinchorro Reef. This was a small group of enthusiastic scuba divers from Mexico City looking for shipwrecks. By chance they stayed at Akumal where CEDAM was having its annual meeting. Most of the group were young professionals led by an architect, Roman Rivera Torres. When they heard about CEDAM of Mexico, they were eager to join and CEDAM was eager to have them, so eager, in fact, that Pablo and the others came to a major decision. They decided to turn the future of CEDAM of Mexico over to this young group, none of whom was over twenty-nine. Roman was elected president of CEDAM of Mexico.

In the September 1977 issue of *Geografia Universal,* a magazine published in Mexico City in the same format as *National*

Geographic, Roman wrote the lead story entitled "Chinchorro: Reef of Shipwrecks and Pirates." He and his companions had been converted to amateur nautical archaeologists. The article listed eighteen ships they had explored on Chinchorro Reef.

Knowing one another, being in the same social set, all living in the same city, it was relatively easy for this young group to line up explorations during which they took along their girl friends, guitars, and wine, and made themselves very much at home with excellent seafood on the barren reef between forays underwater searching for wrecks.

Not that it was all play. These were well-educated, very serious-minded and talented divers who also had good connections with the top Mexican authorities responsible for giving out exploration permits.

25

In 1976 Andy Rechnitzer resigned as president of CEDAM International, to be replaced by Bernarr MacFadden, of Dallas, Texas, a World War II navy fighter pilot and since then a successful businessman. Michael J. Kelly, a Chicago businessman, was named chairman of the board.

Rechnitzer, faced with increasing demands of his job with the oceanographer of the U.S. Navy, said he could no longer handle being head of CEDAM International, where he had "a sense of loneliness and frustration." With no paid staff he found himself doing all the administrative work as well as trying to put out the quarterly bulletin. Members had not been responsive in contributing material for this publication. The chapters were fading for lack of leadership and interesting field activities.

Before relinquishing the reins, Andy pushed a central goal of the organization before the Seventh Annual Conference of the National Association of Underwater Instructors (NAUI) meeting in Miami. Andy had been one of the earliest scuba instructors licensed by NAUI.

He told the group he thought there should be an addition to the "NAUI Code of Ethics" regarding the preservation of underwater wrecks and archaeological sites. As scuba diving had spread, it had become almost traditional for divers to desecrate the wrecks they discovered. Instead of scavenging, the diving community "should strive to protect its mutual underwater assets against damage, depredation, theft, and unnecessary regulatory restrictions."

The public had come to accept the rule "Don't pick the wildflowers" as being mutually beneficial. Such thinking had to be applied to artifacts found under the sea. He said: "Effort expended today to arrest depletion of underwater archaeology and maritime history will provide an enduring source of pleasure and value to all concerned, today and tomorrow."

The diving community had to take the responsibility or through various state laws they would be deprived of a role. Most laws, he said, though their overall intention was good, were "only a noble attempt to realize a satisfactory level of conservation." Despite this, commercial promoters of visits to shipwrecks were flourishing and a black market in artifacts had developed.

There weren't enough professional underwater archaeologists to investigate these sites. Did that mean they should be left unvisited and untouched indefinitely? No, responsible diving organizations could assist in these investigations. And this should be backed by a diver's code to help police wreck sites. It should be an "unwritten law" not to "pluck" these wrecks but to conserve them for the benefit of all divers, present and future.

Over the coming years Andy was to see this ethic gradually catching on, gaining acceptance, though everyone knew that greed would never be eliminated.

A part of the emerging pattern of tight state control to preserve historic shipwrecks could be seen at this time in Texas. Though long active and deeply interested in its historic past on land, previously Texas had given little thought to archaeological sites beneath the sea. This is what started it:

In 1554 four Spanish ships, the *Santa Maria de Yciar,* the *San Esteban,* the *Espiritu Santo,* and the *San Andres,* loaded with 400 passengers and crew, carrying around $10 million in silver and gold, left Veracruz and headed north up the coast before turning toward Havana, their next port of call.

A gale blew the ships onto Padre Island on the Gulf of Mexico where all but the *San Andres* were grounded. Most of the people were drowned. The *San Andres* made it back to Havana. Subsequent Spanish colonial salvage operations recovered 38,000 pounds of goods but little of the precious metals.

There the wrecks rested and rotted until 1967 when a group of treasure hunters from Gary, Indiana, known as Platero, Inc., went to work on the site of the *Espiritu Santo*. They pulled the trigger. A court order stopped their diving. They were operating on property owned by the State of Texas.

Details of this were told in an article by Tommie Pinkard entitled "Treasure, Ships and Dreams" in the January 1978 issue of *Texas Highways*. First was the battle of the courts. The treasure hunters had taken their artifacts back to Indiana. A court decision in 1969 established they belonged to Texas. That same year the legislature passed the State Antiquities Code which established the Texas Antiquities Committee.

The Platero divers, interested only in treasure, had made no archaeological investigations. Work began again. From 1972 to 1976 the Texas Antiquities Committee did a professional archaeological underwater dig. J. Barto Arnold, an underwater archaeologist, was the project director for excavating two ships. The third had not been found.

Much of what came up from the wrecks was in the form of huge "conglomerates," some weighing up to 300 pounds of artifacts encrusted together, everything from dishes and coins to bits of food, even cockroaches. This material had to be cleaned, but the Texas Archaeological Research Laboratory in Austin had no experience. Over the next few years they developed techniques. Dr. Don Hamilton, in charge of this phase, is now perhaps the leading expert in the United States.

Recovered were gold and silver coins, cannon, anchors, navigational instruments, and many other objects, rescued from the concretation and described by Texas State archaeologist Curtis Tunnell as "some of the earliest antiquities ever recovered in the New World. They tell us about the rich Spanish heritage in the Southwest that is often overlooked." To identify the artifacts a contract was let to the Old Spanish Missions Research Library at Mission San José in San Antonio.

They sent a team to Spain to examine sixteenth-century documents. They brought back copies of more than 1,100, plus photos they had made of cannons, firearms, sea chests, and other sixteenth-century items.

Aside from archaeologists, historians, chemists, photographers, artists, writers, and others worked to put the artifacts before the public with a book, a film, and an exhibit. During 1978-1979 this striking exhibit of items recovered from the sea, where they had been for 423 years, went on a tour of Texas cities. Smaller traveling exhibits circulated in schools.

In Texas, nautical archaeology had come of age in one decade. Following the successful excavation of the *San Esteban* and the *Espiritu Santo,* in 1975 J. Barto Arnold and Carl J. Clausen went looking for the third ship, the *Santa Maria de Yciar,* and other shipwrecks. Earlier it had been thought this ship was lost forever with the dredging of the Mansfield Cut through Padre Island.

Using a proton magnetometer they made one survey of ten miles in length at the Mansfield Cut at Padre Island, and the other a fifteen-mile trip near the mouth of the Rio Grande at Paso Brazos Santiago. They reported: "In one five-mile section of shoreline along the Gulf of Mexico we plotted 31 anomalies; two shipwreck sites, three more probable shipwrecks, four possible sites and 22 more anomalies possibly caused by scattered wreckage of ferrous debris."

Their data went through three stages of automated processing with the anomalies plotted for ease of future location and more detailed investigation. The research would go on for years, and it did not appear likely any amateur divers or their organizations would be involved.

In June 1976 CEDAM International followed what was now its usual pattern. It scheduled expeditions to investigate wreck sites on Scorpion Reef as well as exploring the coast and lakes in concert with CEDAM of Mexico. These investigations were followed by the annual meeting at Akumal where the expeditionaries could report their findings to the others who had not taken part. They also heard news of other members, for example, that their advisor "Doc" Edgerton, had just returned from an expedition to the Aegean Sea with Jacques Cousteau aboard the *Calypso*.

Being in an area where the Maya civilization once flourished, it was not unusual for CEDAM International members at these meetings to sometimes divert their attention from the sea to new discoveries on the land, especially if it dealt with Chichén-Itzá—the case now—where three times they had investigated the Sacred Well. The new discovery, made by a CEDAM member, Luis E. Arochi, a Mexican lawyer, involved the main pyramid there.

The Spanish priest, Diego de Landa, who visited Chichén-Itzá in the sixteenth century, was told the pyramid was named Kukulcan after the plumed serpent. But in 1527 when Francisco de Montejo, conqueror of Yucatan, visited the place he decided to call it the "Castillo" or castle, a name it has born ever since.

Built around the twelfth century, the pyramid was nearly 80 feet high, running some 180 feet on each side. A staircase from the top had a four-and-a-half-foot-high carving of a serpent's head at the base. Arochi, visiting the pyramid one day, noticed something about it. It was to obsess him and nearly make him a poor man as he investigated the phenomenon.

He found the Pyramid of Kukulcan was so placed that on the days of the equinox in March and September, when spring and fall begin, something strange happened. At dusk the light from the sun cast a pattern on the pyramid. Beginning at the top a wavy pattern of light and shadow would appear at the top, descending along the staircase wall and illuminating at last the serpent's head.

Then, before the sun finally sank, plunging the pyramid in darkness, these wavy patterns of light that had formed a huge snake would retreat back up the pyramid. Clearly the Mayas had a profound knowledge of astronomy.

In reading Bishop Landa, Arochi found he had indicated the Maya solar year of 365 days began with the month of Yaxkin (first sun) and ended with the month of Xul (termination or end) when the serpent appeared in light and shade on the pyramid. Landa had been told on the last day of the year, Kukulcan, the plumed serpent, descended from the heavens to receive services, vigils, and offerings. It was an important festival and people came from far off to celebrate it.

Arochi reasoned from a practical point of view this signal from the sun would tell the people in March it was time to prepare the land for sowing by burning the forest to get ready for the rainy season. In September it would tell them the rainy season was over.

He could imagine the old priests standing high up on the pyramid while the multitude waited below. Offerings, he thought probably would have been put on the head of the serpent. It would undoubtedly have made a tremendous impression on the minds of the people as they watched the sun god descend to alert them to a new season. And what religious theater!

In the beginning there were many nonbelievers in Mexico as the lawyer slowly accumulated his facts and photographs. He has a photograph of the pyramid he took during these early investigations on the day of the spring equinox when perhaps no more than fifteen or twenty tourists are viewing the structure or climbing the stairs, ignorant of the phenomenon about to take place as a result of Mayan mathematics, geometry, astronomy, and religion.

Arochi has another photograph taken at the same spot two years later. By then the event was well known, accepted, and pushed as a tourist attraction. The grounds around the pyramid are covered with thousands of people waiting, just as in bygone days, for the sun god to come down from the pyramid.

Today Luis Arochi is the manager of the land company at Akumal, close enough to Chichén-Itzá to continue his studies and keep CEDAM members briefed on any new findings.

26

In the summer of 1977 CEDAM International set up camp on Cayo Culebra (Snake Island) in the middle of Ascension Bay south of Tulum on the Quintana Roo coast. The small island was thick with mangrove trees, birds, sand fleas, and, at night, mosquitoes. Natives told them in the old days this island had been used as a prison and that no one could escape because of the deadly Yucatan sharks, the worst in the world.

One of the divers, Julian Lindenauer of the American Embassy in Mexico City, was eager to photograph a shark. Things didn't work out like that, so when the expedition was over he went by himself to Isla Mujeres to try the cave of the "sleeping sharks." It had always been believed that sharks had to keep moving underwater to breathe. But here they had been found motionless in this cave, apparently asleep.

Arrangements were made at the dive shop, Buzos de Mexico, and a guide took Julian, a man and his wife, and a single woman from New York—he never got any of their names—out in a small open outboard motorboat to a spot in the choppy sea above the cave. Julian was pleased to be able to see the bottom 65 feet below, clear and sparkling, important because he didn't have a strobe and had lost his light meter. The wife waited in the boat while the guide led Julian and the others down under a natural coral arch past a large barracuda and dozens of colorful tropical fish.

At the entrance to the cave the guide motioned for Julian to follow him. They crossed over a huge boulder and on the other side found what he had come to see—a nine-foot shark resting on the bottom, perhaps asleep. Julian quickly took some pictures and exited the cave, waiting near the entrance to photograph one of the other divers as they came out.

But the water was suddenly murky. The guide's fin had caught the sand, and it rose like a screen just as the woman from New York disappeared into the cave. Suddenly something like a torpedo blasted from the cave, zooming through the screen of sand, coming directly at Julian and barely missing his left shoulder. It was the shark!

Back in the boat the single woman from New York told them she couldn't resist—she had to pet the shark. Julian said, "This obviously scared the hell out of him!" He said he got so mad at the woman for doing this nutty thing he wouldn't talk to her.

Later, because of the dim light in the cave, only one of his pictures came out, the shark poorly defined. Experts told him it was a thresher, a lemon, a bull, or a nurse. Julian said, "Who cares? At least I got my photograph."

In general over the years the interest of divers in the CEDAM International expeditions in fish had been limited to looking, taking underwater photos of these colorful creatures, or providing supplies for the dinner table. But in 1976 for that year's expedition CEDAM added ichthyology to its interests.

Tom and Marilyn Fifield, as Friends of the Milwaukee Museum, and CEDAM International sponsored inclusion of Dr. Richard Spieler, the museum's curator of fishes, Steven E. Yeo, his assistant, and Teresa A. Noeske, a research associate. At Ascension Bay that year they captured two specimens of an inch-long fish that was new to science, *Starksia occidentalis*.

One objective for including these scientists was conservation. With the rapid development in the offing for the virgin coast of Quintana Roo, more knowledge of local marine life and the ecology of the reefs and coastline would contribute toward preservation.

The scientists first donned their scuba gear to examine lagoons close to Akumal where both saltwater and freshwater fish could be found in the brackish water. From this, and later study, they found reproduction of some fish seemed to be enhanced in the brackish water and certain parasites reduced or eliminated. If true, this information could be important in the commercial farming of marine fish, a future possibility for some of these lagoons.

On Ascension Island in 1976 the expeditionaries set up their tents in a small coconut grove now looking for fish as well as shipwrecks. As they had with archaeologists, the divers were working under the direction of professionals in this pursuit. They used spears, nets, seines, hooks, even fish poison, collecting more than 100 species of fish, most new to the Milwaukee Museum's permanent collection. During both the 1977 and 1978 expeditions, Dr. Spieler and Dr. Noeske continued their research with the ultimate goal of doing a book on the marine life of the Caribbean.

At the 1978 annual meeting Dr. Spieler and others discussed the possibility of establishing a research facility at Akumal. He thought this would encourage scientists to bring students to the facility where, in addition to research, courses in subjects like biology, geology, and archaeology could be offered, perhaps for college credit. CEDAM International members could work in support of the research.

Dr. Spieler was especially worried about sewage disposal along the coast, where the porous limestone would allow seepage, if cesspools were built, contaminating the freshwater flowing into the lagoons and to the reefs.

CEDAM International director Tom Fifield was appointed by the board to look into development of the facility. They decided to call it the Pablo Bush Romero Center for Scientific Studies as a living memorial to this Mexican pioneer in underwater archaeology.

Initially, it was decided, they would build an addition to the dive shop for the facility and begin the process of raising funds and getting equipment. In the future it was hoped there would be room for some classrooms, a lab, and storerooms for equipment and artifacts. It would be available to professionals investigating the history of man in Mesoamerica, both on the land and in the sea.

Pablo said a facility like this could help a great deal in CEDAM's underwater investigations over the years ahead. Sometime, he said, we will find a pre-Colombian prize, perhaps even discover the secret of how jade came to the land of the Maya, an enigma ever since the Spaniards first landed in Yucatan.

By Christmas the large, rectangular addition to the dive shop, the Pablo Bush Romero Center for Scientific Studies had been completed—the first step. Though equipment was meager, it was a beginning. It held a miscellany of CEDAM International equipment.

For the annual Christmas party at Akumal that year the workers and their families as usual gathered for a fiesta, food, music, drinking, singing, and fireworks. Late that night when all had finally gone to bed, a young boy came out on the darkened beach with a skyrocket he had found earlier and concealed under his jacket.

He lit a match and put it to the fuse. The skyrocket rose into the air trailing a fiery tail. It landed on the enormous thatched roof of the dining hall. This stood next to the dive shop, also with a thatched roof as were many of the other buildings in the Maya style. In moments there was a fierce blaze, fanned by wind coming in from the sea.

Within a short period the dining hall and all surrounding structures were burned, including the new center.

Ashes covered all.

27

In December 1977 CEDAM got "on the map."

It was a *National Geographic* map entitled "Colonization and Trade in the New World." The map went with the lead article that month. It was entitled "Reach for the New World," written by CEDAM International honorary advisor Mendel Peterson, now retired from the Smithsonian Institution as its underwater expert.

He was also the principal consultant for the map. On it were famous shipwrecks up and down the coast of North America and in the Caribbean. The only wreck shown along the coast of Mexico or Central America was *Matanceros,* the "Five and Dime" wreck, *Nuestra Señora de los Milagros,* which CEDAM had first explored in 1959.

The map was beautifully done. It showed the annual route of the two Spanish fleets. One, called *Terra Firma,* stopped at Cartagena to load gold, emeralds, and pearls and went from there to Portobelo, Panama. Earlier Inca gold had been brought to Portobelo across the isthmus by mule train and barge and stored in readiness for the great fair of barter and general hell-raising when the *Terra Firma* fleet was in.

The other, the *New Spain* fleet, went to Veracruz. On the other side of the country, galleons had crossed 10,000 miles of the Pacific to bring oriental silks and porcelains from Manila to Acapulco. These treasures were then taken by mule train across Mexico to the port of Veracruz. Here the two fleets rendezvoused and sailed for home, via Havana, facing the threats of storms, pirates, and privateers.

On the reverse side of the map was a handsome cutaway drawing of an armed Spanish galleon of the period. This was surrounded by drawings of a cross section of artifacts recovered from the sea, weapons, shipboard articles, and cargo items bound to and from the New World. CEDAM divers had found samples of these in every category, and some of these were shown in photographs accompanying the article.

On the map the editors noted: "How many wrecks have yet to tell their tales? Modern diving equipment is less than thirty-five years old, and the Atlantic waters Europeans sailed to the Americas cover millions of square miles. We have barely scratched the undersea surface."

One small scratch was made by a young Mexican boy named Manuel Palanco. He came from a family of fishermen. He first began fishing on Chinchorro Reef off the Quintana Roo coast in the early 1940s. On one of his first trips, diving for lobster, he found a wreck. As he grew older, he acquired a snorkel mask and fins as an aid in catching lobsters. He found other wrecks. In 1976 Manuel and CEDAM came together. He became both a member and a guide.

The most promising of Manuel's finds was the main target for CEDAM's 1977 expedition to what they came to call the Forty Cannon Wreck.

In past expeditions the Mexican group usually took the first two weeks and the Americans the second two weeks with some intermixture and overlap between the two groups. This year, because, among other things, of Joe Kelly Hughes' fluent Spanish and friendship with Roman Rivera Torres, the two leaders decided to work their teams together.

There was a total of forty-three expeditionaries and ten crewmen on two boats. The Mexican team was on a boat called the *Alvaro Obregon,* the American team, which included some Canadians, on the *Nojoch Cutz,* which they soon christened, "No Hope."

The Mexicans had the advantage of friendship and past experience working together as a team. The Americans and Canadians had to get to know one another. On a small boat it didn't take long. There were eighteen of them. They drew lots for the sleeping spaces, four bunks in the cabin, six sleeping spaces on the cabin floor, the rest on deck under a ragged tarp. Tasks were delineated and divided. In the crowded quarters all hands had to learn quickly to stow their personal gear and keep things shipshape.

Manuel Palanco, the fisherman, had been hired to guide them to the wreck. Chinchorro Reef covered 200 square kilometers of very tricky water. The two boats anchored off the outer rim of the reef a few miles from Cayo Norte, a small uninhabited island inside the reef. Manuel pointed out the site. Several divers went in work boats over the reef and into the water. Two large anchors were sighted at a depth of fifteen feet. This was it, the Forty Cannon Wreck.

Both teams had agreed to use nautical archaeological techniques to construct a topographical map of the site measuring the relative distances of the cannon to selected base lines. The total area to be mapped was 900 square meters. The cannon and other large objects would be left *in situ.*

Rope anchored to the bottom formed the grid system. The rope lines were laid out on a north-south and east-west axis. Teams of two divers each were assigned to specific sectors of the grid to mark the number of cannon, angles within the grid structure, measure the length, width, and bore, and take photographs. They also made recommendations on the best sectors for further excavation. Each day a dive master kept a head count on those in and out of the water.

By the end of the first day the topographic study showed data on thirty-nine cannon and two anchors. A third anchor was too far out to be included on the grid. Location of the fortieth cannon was not determined. In the selected sectors divers used their hands to fan away sand, bringing up various small objects. The propeller of an outboard motor was fastened to a tubular device running to the bottom with the prop wash removing enough sand to reveal some of the ship's ribs and what looked like the keel section.

Sitting around on deck at night the teams formulated a theory of the wreck. They guessed it was an English ship of the line, something between a brigantine and a large sloop. They judged it had been wrecked sometime before 1850 (the usual problem, a hurricane). Anchors had been dropped to no avail. The ship had been driven on the reef and as it broke up, cannon and ballast were scattered.

There was a second wreck site they wanted to look at in the south. En route the teams could see the rusting hulks of modern ships, relics of past disasters, caught on the reef. One of these was the *Glenview,* a British freighter sitting high on the reef, driven there by a 1963 storm. She was rusting and rotting. A lone ladder hung from *Glenview*'s deck. It was decided to pay a visit to the hulk. They gave up this exploration when Dick Chamberlain, of Houston, Texas, injured his leg when part of the rusted deck gave way where he was standing.

The second site had been listed as wreck no 13. The two boats anchored off Cayo Lobos, southernmost of the reef's three islands. The divers used steel bars to excavate. The work, because of the heavy surge, was hazardous, exhausting, and not very rewarding. Bars of ballast and hollow shot were brought up. Lack of cannon caused the divers to speculate the ship was possibly a nineteenth-century English cargo carrier trading in hard wood.

This brought an end to their investigations. Back in Akumal, where the annual CEDAM International meeting was about to get underway, the expeditionaries reported their findings and showed their artifacts. Before they broke up the American-Canadian team evaluated results. The Mexican group was younger than they, but this didn't make any real difference. They had to admit to language and cultural problems.

The cultural problem mainly involved time. The northern group liked to be up and around early in the morning and get into the sack at night before it was too late. The Mexican group, in the tradition of their country, liked to stay up much later at night and sleep until midmorning. At one point Chris Roberts, a teen-aged diver from Centralia, Illinois, went aboard the *Obregon* early in the morning and blew his bugle. The Mexican divers tossed him over the side and went back to sleep.

Noise had been a problem. Scuba needs air tanks and these had to be refilled almost continuously from the air compressors. These made noise. Besides, there was the noise from the boat's engine while underway, and noise from the bilge pump when the boat was at anchor.

Rain interfered with sleep. It would be better if there were more bunks for tired divers. A short prior course in underwater archaeology would be helpful. But they recognized little could be done about the noise or bunks. It had been a grand adventure. Most were determined to come back next year and continue the exploration.

CEDAM International's annual meeting was the usual collection of diverse topics. Kenneth Hollingshead, president of the Washington, D.C., chapter took them from the warm, clear waters of the Mexican Caribbean to the dark, cold waters of the Mullica River at Chestnut Neck, New Jersey, via a lecture and slide presentation.

Hollingshead, a U.S. government oceanographer, had been working for several years with a team of divers in these inhospitable waters, where the visibility was about one foot, to investigate some British merchant ships, captured and sunk there by American privateers in 1778.

Board member Adm. Howard D. Greer, USN, commander of the Naval Air Force, U.S. Atlantic Fleet, told how the U.S. Navy the previous year had recovered an F-14 jet fighter weighing twenty-five tons from a depth of 1,900 feet in the North Atlantic. Due to brake failure, the multimillion-dollar plane had rolled overboard from an aircraft carrier and sunk in the North Atlantic during NATO exercises.

Part of the tension of the complicated deep-sea rescue effort had been to prevent the Russians from getting their hands or hooks on the plane and its secrets. Fighting thirty-foot high waves and winds up to forty miles per hour, three times the navy ships got lines around the plane, each time to have the cable break and the plane settle again to the bottom. The fourth attempt was successful.

Members noted with sadness the passing of honorary board member Wernher von Braun, the famous rocket expert, as responsible as any for getting a man on the moon.

More Mayan mysteries were unveiled. Members were told of the color, Maya blue, found nowhere else in the world. Its bright color was one more mystery. No chemist had been able to determine what was responsible for the color, what the dye was, and what made it so incredibly stable.

Dr. Luis Gonzales Calderon, a Mexican physician and archaeologist, told them of his theory on the origins of the Olmecs. He had just returned from a trip to the Peoples Republic of China where he had studied similarities between the Chinese and the Olmec. He claimed the origin of the Olmec Indians could no longer be considered a mystery. They had come from Asia.

For thirty years he had practiced medicine in Coatzacoalcos, Veracruz, where he had a private hospital. Knowing of his interest in archaeology, many of his patients had sent him gifts of clay figures from the two main Olmec centers, San Lorenzo and La Venta. Studying these heads he had noted facial characteristics that seemed pure Chinese.

His research dated the Olmec back to 1200 B.C. During his trip to China he found the Chinese had dated similar artifacts to 2000 B.C. Mainly, he found, the Olmec heads, hairstyles, eye configuration, and mouths looked exactly like those he found in China. He had just published a book on this theory.

It was possible, he thought, the Chinese had been shipwrecked on the American continent as part of their early explorations. He noted only the Chinese and the Olmec-Maya used jade in their culture. Jade was not indigenous on this side of the water. He also believed the Phoenicians were involved in the evolution of the Olmec civilization.

Gertrude Blom, who turned seventy-seven at the meeting,

brought members up-to-date on the Lacandonian Indians with whom she had spent fifty years of her life.

On the one hand they were well off, even rich compared to other Indian groups in Mexico. Income came mainly from preparing mahogany trees for market. But this had brought about a demise in their life-style. Other Indians had invaded the jungle to share in this profitable business. Radios, roads, and airplanes were opening up their country to exploitation. Unless someone stepped in, the jungle would be destroyed by mahogany seekers. The jungle did not contain sufficient life-support systems to maintain the huge population that had infiltrated it.

Bernarr MacFadden, CEDAM International's new president, closed the meeting declaring his goal for the coming year was to increase membership* by several thousand. He hoped most would be in the 19 to 30 age bracket as the current average age of members was over 40.

Next year, he said, they would return to the Forty Cannon Wreck.

One more scratch.

*Those desiring to join CEDAM International may do so by sending $25 for initiation fee, and $15 for annual dues to: CEDAM International Membership Services, 436 Monssen Dr., Dallas, Texas 75224.

28

Early in 1978 I got to thinking about CEDAM International's expeditions over the years. A weakness was that they had never had a professional nautical archaeologist in charge of any of their underwater digs. Terrestrial archaeologists had been at the *cenote* excavations. I suggested to CEDAM International's president, Bernarr MacFadden, that we discuss this with Dr. George F. Bass, president of the Institute of Nautical Archaeology (INA) at Texas A & M University, College Station, Texas. He was perhaps the nation's leading nautical archaeologist and, besides, was one of CEDAM International's honorary advisors.

Unfortunately, we couldn't see Dr. Bass when we made our visit at the end of May. He had gone back to continue work on the excavation of a thirteenth-century Byzantine ship near Marmaris, Turkey. This was the kind of continuous, painstaking work I had in mind for one of our interesting wrecks off the coast of Quintana Roo. Instead we talked with Dr. Frederick H. van Doorninck, Jr., himself distinguished in the field and about to leave for Turkey; J. Richard Steffy, an expert on the reconstruction of ship hulls; Roger C. Smith, a graduate student—turned out I knew his father; and later with Dr. Don Hamilton, the wizard at artifact preservation. I also talked with Dr. J. Barto Arnold, in charge of underwater archaeology for the Texas Antiquities Committee.

It was an eye-opener. When it came to pure nautical archaeology, not general surveys or explorations of wreck sites, professional underwater excavations could almost be counted on two hands. The science was really in its infancy. For one thing there simply wasn't enough nautical archaeologists to go around. Besides this, it took time and money.

Andy Rechnitzer said he was with George Bass on his first dive. It was in the Western Hemisphere, a shipwreck off North Carolina. Dr. Bass and others, we were told at the institute, had excavated, between 1961 and 1964, a seventh-century Byzantine wreck resting on a slope of the Yassi Ada Reef, Turkey, at a depth of 120 feet. Between 1967 and 1969 they had excavated a fourth-century Roman ship wrecked nearby on the same reef. Now they were working on the Marmaris wreck. It had been carrying a load of glass when it went down in 110 feet of water. That was pretty much the Mediterranean.

Work was going on to excavate a seventeenth-century shipwreck in Mombasa harbor, Kenya, Africa. At Penobscot Bay, Maine, work was continuing on a careful excavation of the remains of the American privateer *Defense,* scuttled during the Revolutionary War. Donald Keith, a doctoral candidate, was working in Korea on the wreck of a fourteenth-century junk, and there were some other projects, but not many.

Most of the watery world hardly had been touched.

Texas A & M had one of the few facilities, a converted World War II factory, anywhere in the country, or the world, for that matter, devoted to the conservation of artifacts recovered from the sea. Work was sent there from all over the country.

The Institute of Nautical Archaeology was housed in a small one-story wooden building, also of World War II vintage, some distance from the Texas A & M campus at College Station. Though unpretentious, in nautical archaeology this was where it was at.

Dr. Bass had formed the Institute and developed a network of supporting institutions around the country. There had not been many graduates. It was a tough road. For example, back in 1971 Roger Smith, whom we met, had been a senior at the University of Virginia. He asked Bass about a future in underwater archaeology and was told not only was it a long academic grind, but also there were few opportunities to practice the skills. Smith spent the next seven years getting ready. In the coming summer he was to lead a team directing INA's first probe in the Caribbean. He was to make an inventory of shipwreck sites in the Cayman Islands as the basis for his master's thesis research.

Smith said, "For me, nautical archaeology has all the elements of intellectual and physical pursuit combined into one. It is an academic synthesis that allows me to work at reconstructing one part of human history while at the same time be outdoors and experience nature on and under the ocean."

Bass also looked south the following year when he and Don Keith, who had worked on the *Monitor,* went to Mexico to conduct a short course in underwater archaeological field methods at the invitation of the archaeological diving group of the National School of Anthropology and History. They conducted training dives at the Lake of the Half Moon in the State of San Luis Potosi. Pilar Luna, leader of the Mexican group, had worked out of Akumal in 1975 with CEDAM International and the Dr. Nancy Farriss expedition.

At the lake, where in the ancient past thousands of terra-cotta figures had been placed, Bass decided on a novel approach. Rather than search for real artifacts he had the students develop their own artificial site sprinkled with modern pottery broken into sherds. Over the years sports divers had recovered thousands of the small figures from the lake without, of course, keeping any plan of where they were found.

Bass has little use for amateur archaeologists. After this trip he wrote in the INA newsletter: "Mexico's potential for underwater research is enormous: *cenote*s and springs associated with pre-Colombian civilizations, inundated prehistoric sites and shipwrecks on both the Caribbean and Pacific coasts exist in abundance. But the predations of amateur divers and professional looters are taking their toll."

The main thing was we accomplished what we set out to do. Jack B. Irion, twenty-five, a bright, young graduate student at the Institute of Nautical Archaeology, agreed to supervise further excavation of the Forty Cannon Wreck in the coming summer.

His first field excavation had been on land six years earlier, a thirteenth-century castle at Hadleigh, England. That same summer he dove on the Spanish caravel off the coast of Texas as part of the Texas Antiquities Committee team. He had worked with Dr. Bass on the Yassi Ada wreck in Turkey. His wife had gone with him there for their honeymoon.

Jack had worked on a Greek sanctuary at Metaponto, Italy, and, to vary things a bit, on a stagecoach depot at Anderson, Texas. He had obtained his B.A. in classical archaeology from the University of Texas at Austin in 1974, his M.A. in 1977, and was working at INA as a special graduate student in nautical archaeology. He was past president of the University of Texas Anthropological Society and was currently teaching a course at the university on "The History of Seafaring and Nautical Archaeology."

Along with this he was taking courses in Spanish, and Meso-american Etruscan archaeology. We were glad to get him.

Within the year he was made CEDAM International's director of education.

29

A visitor to the land of the Maya in the late 1970s would find many changes had taken place in just a few years. It was said Mayan women had to make a pilgrimage to Cozumel Island to be assured of eternal life. The island now had a growing number of hotels to accommodate the flood of tourists, some 180,000 annually, of which nearly 20,000 were scuba divers. Planes and boats took tourists on side trips to the mainland—all a far cry from how CEDAM found life on this remote island two decades earlier.

Then there was the new city of Cancún. Here tourists could sample the pleasures of a beach resort in the tropics or take trips to Chichén-Itzá and other historic spots on the Yucatan peninsula. It was only an hour by taxi from the airport south to Tulum, Xelhá, and Akumal. The Mexican government had mounted an energetic promotion for Cancún and the surrounding area and the tourists were pouring in. It was increasingly common to see busloads of Germans or Japanese over on a charter flight.

Still, Akumal had escaped. A visitor there in early 1978, seeking to escape the rigors of winter up north, after sleeping on one of the fairly hard beds in one of the huts of the Villa Maya on the grounds of Hotel Akumal, would know he was a long way from home.

He would look up from his bed to a cathedral ceiling of bare poles lashed together and covered with thatch. An overhead fan could stir the air, but there was no need as a fresh sea breeze came through the windows, the only air-conditioning. Lying there he could see the sun slanting in, hear the crash of the surf and the sound of tropical birds chattering in the coconut trees.

Inside the hut were open concrete shelves to hold luggage, clothes, and other articles. There was no telephone, radio, newspaper, or TV. No pictures hung on the bare walls. Furniture was plain and minimal, twin beds, a round table, three straight-backed chairs. Electricity was generated on the grounds. Light was supplied by a low wattage bare bulb hanging from the ceiling. There was a bathroom and a shower. The toilet gave strange moans when flushed.

These huts were in clusters, following in design and construction that of the Mayas. They were made of blocks of limestone plastered over. Windows were louvered. They were situated in the coconut grove that ran along the shore. On the beach nearby the sand was like powdered sugar, the water blue-green. In the distance, with a line of surf, was the barrier reef. Very few people were ever on the beach.

There was a huge, circular thatch-roofed dining hall, the dive shop, a small gift shop, the hotel office, CEDAM's museum, five or six one-story condos, some bungalows, and further up the beach several private homes. Pablo Bush had built a house on the point. Some of the museum's cannon were mounted on the seawalk there where visitors could view them. In Pablo's living room was a painting of the beach when he had first seen it years ago with just a fisherman's lean-to under the palms. On the rocks jutting out into the sea was a small statue of Our Lady of Guadalupe quietly regarding the sea.

Now and then a group would visit Akumal like that of the Audubon Tour, birding in Quintana Roo, Chiapas, and a part of Tabasco. They saw 227 different species of birds. But this was not often. The main crowd came from tour groups, Japanese, German, Canadian, American, and others, stopping by for lunch after seeing Mayan ruins.

One visitor to Akumal about this time was M. Timothy O'Keefe, Maitland, Florida, editor of the *International Diver's Guide,* checking out diving locales. In writing of this he said Akumal was "slumbering," but "if properly promoted it could become one of the hottest diving areas of the next decade."

Unlike many places in the Caribbean he found the Mayan natives happy, honest, and helpful. Lacking TV, radio, and newspapers, time became unimportant. Akumal was a special mood. Then, there were all the ruins. But for divers, it was heaven. These underwater scenes always made divers who wrote of them search for poetic ways to express themselves.

"What a reef!" he wrote. "Soft corals and sea fans, predominantly white in color, crowded one another for space . . . it was like floating above a forest of bonsai plants freshly painted with snow. For some reason thoughts of Christmas trees kept recurring as did nighttime walks in a forest with a moon so bright it bathed everything in silver."

Then there were "the magnificent stands of elkhorn coral, bent all in one direction like a tree eternally blown by the wind; giant gorgonian fans clustered in a line and standing straight like finely laced tennis rackets . . . I wanted to go crazy with my camera."

O'Keefe wondered which way Akumal would go—on the one hand to concentrate promotion to diving groups interested in research and study, or on the other to hosting a small but affluent diving clientele. Neither approach was quite fair to the diver with limited money who wanted to see these underwater wonders. So far, he thought Akumal had maintained a good equilibrium, but could it continue in the future? He wasn't sure.

"Hopefully," he said, "hordes of divers will never find their way here, delivered to the reefs in the cattle-car boats carrying thirty to fifty people as in the Florida Keys and some parts of the Caribbean."

Who could say? Early in 1978 not too far down the coast a visitor with luck in picking the right spot could dig in the sand for half an hour and come up with ten or fifteen glass beads washed up from the Matanceros wreck site or put on a dive mask or scuba and not far from the shore still perhaps find artifacts among the coral.

But times were changing and fast. Oil globules from passing tankers made a dark and dirty line along this unprotected shore, and, here and there, beer cans and other debris had been left by those who did not care.

30

The two local veteran fishing boats, *Alvaro Obregon* and *Nojoch Cutz,* were again chartered for CEDAM's 1978 expedition to Chinchorro Reef, but this year there was an important addition. This was a spanking new eighty-foot yacht, the *Sixceas,* owned by "Adm." Harold E. Crane, of St. Petersburg, Florida, a manufacturer who had designed and built the boat himself in a factory ordinarily making doors.

Sixceas, on her shakedown cruise, was loaded with luxuries, ice-makers, stereo, air-conditioning, and lots of room both below and above decks. She had 1,000 HP engines, weighed 100 tons, and could do 20 knots. It was intended to use the boat as a support facility working with the other two boats. In practice *Sixceas'* comforts were too strong a lure, and it wound up as both a sleeping and dining facility for many of the divers.

The expedition was divided into two phases, the first with fifteen members for a week, the second with forty-one members for two weeks. Jack Irion was to supervise further excavation of the Forty Cannon Wreck.

Main task of the first group was to install a grid system of pipe over the wreck and buoy major markers such as certain anchors and cannon. This was a serious effort. Workdays began at 5:30 A.M.

One important thing had been cleared up the year before. That was to establish this really was a wreck site and not just a place on the reef where cannon and anchors had been dumped in order to lighten a ship. Location of the wooden ribs of the ship had proved it a true site.

By the end of the first day the grid was in place. All had gone very well. From the reference points it could be seen the wreck was perpendicular to the reef with large anchors lying to the northwest and stern anchors in shallower water to the southeast.

Ships in the old days, using rope lines on their anchors, had these rapidly sawed through on the jagged reefs. Anchor chains might have held.

The divers made a discouraging discovery. Part of the wreck had been heavily dynamited at some time. Artifacts recovered from this section were in poor condition, shattered pottery and crushed wood planking.

Irion set some of the divers to scraping and chipping the encrustration from the cannons to search for identification marks. On the second day the letter *E* was located on the trunnions of three cannon. Over the next few days further scrapings revealed the letters *F* and *G* on several other cannon. Divers were discouraged when no further marks, numbers, or dates could be found on the encrusted weapons.

In one section of the grid divers found a heap of bones, thought at first to be those of some of the crew. These turned out to be animal bones, remarkably free of encrustration after so many years under the sea.

Efforts were made to trace the ribs of the ship to the keel. Was this the ship or just a piece of it that had broken off when the ship foundered? Eventually they uncovered what they thought to be the keel. It was obvious that dynamiting had destroyed much of the planking.

Though the expeditionaries recovered artifacts like grape shot, glass and pottery fragments, a cannon loader, and other items, the yield was disappointing. They speculated much of the ship's fittings had been salvaged by the dynamiters.

Irion suggested opening up a new work trench in an area

where ballast had been found. This proved productive. More frames and planking were uncovered.

The second phase of the excavation was accompanied by intermittent bad weather and other problems. It would seem that two constant problems of these underwater expeditions would be cranky air compressors to fill the dive tanks, and cranky cooks to fill their bellies. "Admiral" Crane's wife, Dolores, still getting used to their new boat and life at sea, volunteered to handle cooking on the *Sixceas*. She hadn't reckoned on the continuing stream of famished divers and was almost undone by the experience.

First there was the sun, blazing hot. Several of the Americans came down with sun poisoning and heat exhaustion. Then one night a squall came up with winds gusting up to 50 knots. It afforded a glimpse of wrecks of the past. The anchor on the *Sixceas* wouldn't hold, and the boat was blown fifty yards toward the reef before the storm subsided.

The ichthyologists from the Milwaukee Public Museum continued their study of marine life on the reef, while other divers with spears made their own study, going after fish and lobsters for the cooking pot.

Artifacts continued to be recovered from the wreck site—cups, bowls, an olive jar, iron pieces, nails, a glass bottle neck, a sword, and a ceramic plate with a design before the expedition was over.

Back at Akumal, where the annual meeting of CEDAM International was soon to take place, the members of the expedition—the nautical archaeologist, students, marine scientists, ichthyologists, lawyers, doctors, dentists, architects, photographers, and businessmen—gathered for a postmortem. Mendel Peterson looked at their "goodies," examined their underwater drawings of marks on the cannon, and heard their oral reports.

This expert on the identification of cannon and other artifacts recovered from the sea confessed he was baffled. Based on his preliminary examination he said the ship might be either Spanish or Dutch, the latter possibility emerging from the discovery of basalt ballast. It was a mystery.

From here on the team began to refer to their find as the "mystery ship" in addition to its designation as the Forty Cannon Wreck.

The clues were really very meager. Jack Irion studied them and conducted further research over the next year. He wrote his findings. The vessel was most likely a warship of the frigate class, relatively large (150 feet) and fast, the kind of ship used for convoy or privateering.

Earlier, he said, these frigates carried from 24 to 28 cannon, but toward the end of the eighteenth century were armed with 40 guns or more. "With 40 cast iron cannons, ranging from 12 to 18 pounders (the weight of the shot) our vessel clearly fits into the eighteenth-century category," he wrote.

And what of these cannon? Foundry markings of E, F, and G had been found on the right trunnion of some of the cannon. This was an early form of manufacturer's trademark. Irion was unable through his research to find manufacturers for the E- and F-type cannons.

He did find an identical symbol of the G-type symbol in Mendel Peterson's authoritative book, *History Under the Sea—A Handbook for Underwater Exploration.* This appeared on the trunnion of a cannon recovered from the York River in Virginia. It was attributed to the foundry of Graham and Sons, circa 1782.

But, Irion added, the fact that this particular cannon was English didn't mean CEDAM International had found a British vessel. England was a center for the manufacture of cannon and they were widely exported to other nations. If the cannon had belonged to the Royal Navy there would have been the "British Broad Arrow" symbol.

So the nationality was still a mystery.

Divers had brought up examples of two types of shot fired from the cannon, bar shot and varieties of lead balls, some clustered into grape shot. Bar shot looked like a dumbbell, with an iron rod connecting two flat discs. The shot furnished no clues.

Through his detective work Irion was able to pin the date of the wreck to within twenty years, between 1780 and 1800. He based this not only on the cannon, their size, design and number, but on the shape of a glass bottle neck and some of the pottery found on the site.

The bottle neck was hand-blown, from a liquor bottle of dark green glass. Among other details he had noted two extra layers of glass had been "laid on" around the mouth of the bottle to provide it with a lip for securing the cork stopper with wire.

Irion divided the pottery sherds into two types: one from common jars used to store anything from water to olive oil, the other represented by a single small sherd, possibly from a plate. It had a blue-on-white floral design. This was known as Majolica ware with a long history in both Spain and Mexico.

Judging from the three-meter-square section of the ship's hull that had been cleared, Irion believed a lot more of the ship's hull lay buried beneath the sand and ballast stone. Clearly much more scientific excavation and research needed to be done to unravel the mystery.

At this writing this has not been done. The Forty Cannon Wreck still holds her secrets.

31

Each year editor Larry Booda, himself a scuba diver and member of CEDAM International, devotes one issue of his magazine, *Sea Technology,* to an overview of all aspects of diving. The emphasis, of course, is on new technology as nations gain momentum in probing the depths of the sea for oil, minerals, food, and defense.

In January 1978 he devoted a few lines to CEDAM International, saying it was a group of "elite divers." At the end of the year CEDAM of Mexico held a meeting at a mountain resort near Cuernavaca, in which CEDAM International joined to celebrate the twentieth anniversary of this "elite" group.

The question was, looking ahead for the next twenty years, where were the organizations heading?

Meeting in the Hotel Hacienda Cocoyoc in the State of Morelos, a historical hacienda that once belonged to Hernando Cortes, participants found a varied program. It was hard to tell about the future.

There was a historical lecture by Jesus Bracamontes, Mexico's top expert on ship building in the sixteenth century. Owner of the most complete archive on the subject, he was also a model ship builder par excellence. He used one of these, a galleon, to point out common clues to identify a shipwreck of this era.

Engineer Oscar Constandse told members human factors were more important than technology or equipment. "We may not have our own research ship," he said, "and our compressor may not be the top of the line, but we do have a group of active, generous, capable, professional, and enthusiastic people."

Pablo Bush Romero read the names of CEDAM's original founders and their objectives twenty years earlier when the organization had been created. At the meeting, with typical Mexican hospitality, there was much food and drink, fireworks, exhibits, music, dancing, and even an Andalusian horseshow.

But in all its aspects perhaps the most important was political. The government of Mexico, with representation from seven ministries, had set up the Intersecretarial Commission for Oceanographic Investigation (CIIO) for combined efforts involving exploration and research in the sea.

Alfonso Romero-Erazo, an industrialist and former president of CEDAM, made a speech in which he spoke of the importance of an alliance between CEDAM and the government's Intersecretarial Commission for Oceanographic Investigation. The representative for Adm. José Manuel Montejo Sierra, head of CIIO, enthusiastically accepted CEDAM's offer and promised full cooperation in the future.

Looking ahead this was one of the strands, and there were others.

CEDAM International had dissolved its chapters. They just weren't viable. The Canadian ambassador to Mexico, James C. Langley, had been interested in forming a Canadian CEDAM, but later he moved on and this didn't work out. The Scottish Sub-Aqua Club had been interested in getting involved, but nothing had come of this.

Mendel Peterson had pointed out to me that over the years almost every major shipwreck anywhere in the world had first been discovered by sponge divers, fishermen, or treasure hunters. This was continuing to happen. Clam boats operating off Ocean City, Maryland, had earlier in the year detected an obstruction on the seabed with their electronic equipment.

Michael Freeman, owner of American Watersports Co., Oxon Hill, Maryland, picked up the tip and went looking with his sixty-five-foot salvage boat, the *Buccaneer*. In ninety feet of water, eleven miles off the coast, he found a 420-ton Civil War tug, the

USS *Nina,* upright on the bottom. A storm had sent her down sixty-eight years earlier and a massive search had been unsuccessful.

Freeman and several other divers brought up the anchor, a huge brass bell, hardhat diving gear, a gold watch, the binnacle, and a beautifully crafted three-ton brass heat exchange used to convert steam back to water.

Most of these artifacts were donated to the U.S. Navy Museum in Washington, D.C., where the bathyscaphe *Trieste* was being readied for permanent display, the vehicle which, under Andy Rechnitzer's direction, had plunged to the world's record depth of 35,800 feet twenty years earlier.

The point was that, although shipwrecks would continue to be found, it was likely that most of the new ones would be in deep water where special equipment and training would be required for salvage, even if permits could be obtained. Part-time sports divers probably could not expect to be much engaged with these wrecks.

In an energy-hungry world, millions were being spent on the development of new and improved means to search beneath the ocean. Not only sophisticated instruments operating from space and aboard ships, but a whole new panoply of underwater vehicles had been created to aid in the search for oil and gas.

By the end of 1979 the National Oceanic and Atmospheric Administration reported about 180 remotely operated underwater vehicles were in use or being built around the world to crawl and explore the ocean floor. It said an additional 120 of these vehicles, called ROVs, were being used by various navies to neutralize mines.

Other underwater units to be towed or to "swim" were working for the scientific research community. These vehicles were being used to inspect underwater structures, to assist divers, to bulldoze, to map, to take water samples, and to perform many other functions. The report said that ROVs had increased by 1,000 percent in the past five years.

With all this activity underwater surely more and more shipwrecks would be discovered. Mendel Peterson told me: "I believe within a decade or so in these depths and with the new technology we are going to find a Spanish galleon, virtually intact."

At least professionals in nautical archaeology would benefit from the new technology. In fact, a few years earlier a team from the Institute of Nautical Archaeology had conducted the first nautical archaeological experiment with saturation diving, later calling it nautical archaeology's new frontier.

Target was the *Capistello* shipwreck, a third-century B.C. Greek merchantman. INA had teamed up with the Sub Sea Oil Services (SSOS) for the project. The excavation was used as training for a group of SSOS trainees learning to work beyond the conventional fifty meter scuba limit. This was accomplished by breathing special nitrogen-free mixtures of helium and oxygen —heliox—in what was called saturation diving.

This enabled divers to have a dramatic increase in bottom time as well as working at deeper depths. Decompression took place in a special chamber aboard the SSOS training ship *Corsair*. The nautical archaeologists used a small SSOS two-man submarine to observe work on the bottom and to take photographs. Working at sixty meters—where pressure on the body is almost twenty tons—they found the results were good. This was the deepest excavation ever attempted under archaeological control.

Dr. Donald A. Frey, an INA staff member heading the team, later wrote: "At a time when many shallow wrecks in the Mediterranean have been destroyed by sport divers seeking treasure, deeper wrecks may hold a singular promise for future work in underwater archaeology."

Still there were persistent problems. There weren't enough nautical archaeologists trained for this deep water work, and the divers who were usually weren't trained in archaeology. It was often frustrating, Frey said, trying to direct the work from topside, though the submarine helped. Like the CEDAM International leaders, after the Forty Cannon Wreck first expedition, he felt a short course on excavation procedures for the divers could be helpful.

But those divers were professionals. In the coming twenty years would there be room for divers in organizations like CEDAM International to work with nautical archaeologists on deep water wrecks? Who could say? Perhaps sports divers of the future would be attracted to the deeper depths.

Duke University Scientists had found a way for professional divers to double the practical working depth from 1,000 to 2,132 feet by a new mixture of breathable gases, an astonishing forward development in opening up the ocean in the future.

One way could be through diving clubs buying their own submarine. George Kittredge, an ex-navy submariner, had been producing such in Warren, Maine. His one-man "personal" sub was twelve feet long, had a mechanical claw, a speed of four knots, and could dive to 250 feet. Cost was around $12,000, and over the past eleven years he had sold twenty-nine of his little subs for underwater exploration. Future plans called for development of six-passenger subs especially designed for underwater tours.

Certainly in the future there would be an increasing number of university students diving and working on shipwreck sites under the direction of nautical archaeologists. This would be for college credit. An example of this was the find of at least six shipwrecks from the historic Battle of Yorktown in the near-zero-visibility waters of the York River in Virginia.

Divers of the Virginia Research Center for Archaeology raised a 1,000-pound cannon believed from the flagship of Lord Cornwallis. John Sands, assistant director of the Mariner's Museum at Newport News, Virginia, said this could be the richest depository of eighteenth-century shipwrecks in the country. Work would go on there for years. Mendel Peterson said there were enough wrecks in the Mexican Caribbean to keep CEDAM International busy there for half a century.

Perhaps over the coming years, if CEDAM International divers became more knowledgeable in nautical archaeological excavation techniques, they too would be called on to help the professionals and tension reduced between the two.

Nondiving members might develop skills in the field of

nautical archaeological research. I found it interesting that at Georgetown University, my old school, research of this type had turned up information on two ancient cannon that had guarded the east entrance to the Healy Building since 1898.

Local historians discovered a bill from a used-cannon dealer suing Lord Baltimore, first governor of Maryland, for nonpayment for these weapons. They were used on his ships bringing the original Catholic settlers to Maryland in 1634. An expert confirmed that the cannon were Spanish-made. By deduction, the outfitting of the ships in 1630 with the defeat, two years earlier, of the Spanish Armada on English shores pointed to one conclusion. Where else would an English dealer in used cannon at this period be most likely to get Spanish cannon? From the shipwrecked armada, of course.

Such historical detective work could bring its own pleasure.

More emphasis on conservation could be another avenue toward the future. At CEDAM International's 1979 annual meeting at Akumal, Dr. F. R. Scroggins, president, and Thomas L. Kimball, executive vice-president of the National Wildlife Federation (NWF), urged exactly this—more citizen participation in conservation efforts would be needed in the years ahead.

Kimball urged members to take an active interest in the law of the sea, particularly the mining of minerals on the seabed. He said when this happened a great deal of silt would be developed that could kill living coral beds. Members could also help in such things as the fight for the preservation of the largest animals on earth, the whales, and endangered species of certain birds and fish. Much more needed to be known about the ecology of the sea.

President Scroggins said the NWF, a citizen's organization, had 4.1 million members throughout the world and an annual budget of $29 million. They had an adult magazine and a children's magazine, each with a million circulation. NWF published weekly reports on bills in Congress having anything to do with conservation, fish, and wildlife. The work of NWF and that of CEDAM International, he thought, were very much in line.

It could work, divers involved with professionals in conservation projects, as when the CEDAM International divers on Chinchorro Reef helped the ichthyologists from the Milwaukee Museum to catch, tag, and photograph fish. And it was logical. Andy Rechnitzer said diving was like big game hunting. At first the idea was to kill—animals or fish. Then most moved on to photography. For many, after a time, photography becomes a little boring, he said, and interest turns to underwater conservation or archaeology, both endlessly interesting.

What would the future bring?

32

There was still another route to the future—the search for gold, skyrocketing in value.

When my wife and I were in Spain early in 1979—where I had served as counselor for public affairs in the mid-1960s at the American Embassy during the exciting and successful underwater search for a missing H-bomb—I heard of another treasure in gold beneath the sea.

There were supposed to be 108,000,000 pieces of gold aboard a galleon, *Monmouth,* sunk along with other treasure ships by the British in 1702 in Vigo Bay in Galicia in the rainy north of the country. It was said the Spanish government had offered to share fifty-fifty with anyone who could find it.

Treasure hunters over the decades had criss-crossed the bay, searching without success. Even Jacques Cousteau had had a go in his famous "diving saucer." Rivers pouring into the bay bring a torrent of mud. Generations from various nations have probed this muck without success.

There is no question. The appeal of treasure to be found under the sea has been, is, and will be a magnet drawing men beneath the waves.

Nautical archaeologists have not taken kindly to these treasure hunters. In January 1978 this subject was addressed at a workshop held as part of the Ninth Conference on Underwater Archaeology, held jointly with the Eleventh Annual Meeting of the Society for Historical Archaeology in San Antonio, Texas. The workshop had the ponderous title of "Workshop on Cultural Resources Management Policies and Procedures for Shallow Water Historic Shipwreck Sites on the American Inner Continental Shelf."

Actually it was a formal confrontation between the underwater archaeologists and the professional treasure hunters. The latter were unhappy with state regulations restricting access to marine sites. They also made the point that under the free enterprise system the professional treasure hunter was more capable of doing the job because he could obtain funds for such an excavation more easily.

Reaction of the underwater archaeologists was predictable. J. F. Muche, editor of *Search, The Journal of Undersea Archaeology, Maritime History and Related Fields of Study,* in an editorial review said: "The day of the professional treasure hunter is dead. Knowledge, techniques, and duty to the public trust and heritage should not allow it to be resurrected. We have come too far to stand back and allow the destruction of a cultural heritage in the interest of individual profit. If we allow this to happen, we will surely lose, and deservedly so, the recognition as a science which we have fought so hard to achieve."

Roger Miklos, a professional American treasure hunter who had recovered a fortune in precious metals under the sea, in an interview said treasure hunting was hard, dangerous work. He said over the years he had broken almost every tooth in his mouth getting slammed up against boats by the waves.

He also knew something of history of the ships and their treasure. It wasn't the intrinsic worth of the coins but their historical value, and this had to be proved, especially by reference to documents in the Archives of the Indies in Spain.

Without such documentation he said the historical value of the coins would be almost valueless. For that reason alone he wouldn't try to cheat the State of Florida of its 25 percent tax levied on treasure finds. He needed both the publicity and the state's verification to authenticate the treasure.

Consequently, like a nautical archaeologist, he had his own wide collection of archival records, documents, and books. Clients and collectors had to be convinced.

One of the most famous American treasure hunters was Mel Fisher, who found the Atocha treasure off Key West. Others had been looking for three centuries for *Nuestra Señora de Atocha,* a Spanish galleon that went down in a hurricane in 1622 with a cargo of forty-seven tons of silver and gold. Begun in 1971, the treasure seven years later had cost him $5 million, an investigation by the SEC, a continuing fight with the federal government and the State of Florida, and four lives, including that of his oldest son.

In the fall of 1978 a landmark decision came when a federal judge in Miami decided the treasure he had recovered belonged to him, and none had to be shared with either the federal government or Florida. Earlier, to fund the project, he had sold shares, promising investors one-quarter of one percent share and had raised $1 million before being stopped by the SEC, who said he was dealing in unregistered securities.

As the gold came up over the years Fisher carefully had put aside 25 percent for the Florida tax. But the treasure was tied up and he couldn't make any profit from it. By 1977, in a basement vault of the State Archives Building in Tallahassee, there were some 3,000 Atocha items worth more than $2.3 million.

The U.S. district judge ordered the state to give this treasure back to Fisher. Treasure hunters all over the country rejoiced.

A group called Hatteras, Inc., with a different kind of "treasure," took their case to court in Texas. They had found an early iron steamship, sunk in 1863 by a Confederate raider, sixty feet deep under a layer of sand thirty miles south of Galveston, Texas. They wanted to raise it for public display but had been unable to obtain clearance from either the U.S. Department of Interior or the Texas Antiquities Committee.

The Hatteras, Inc., group said they were bringing the suit so they could keep treasure hunters off the boat, raise it, show it to the public, and get their expenses back. Since the boat had been down there 115 years, the state and federal authorities didn't see the hurry in getting it raised.

In 1979 CEDAM of Mexico for the first time also got into the treasure business. They had worked out an agreement with the Mexican government's Intersecretarial Commission for Oceanographic Investigation on the division of any treasure that might be found. If successful, it was understood the government would invest its share in further exploration of other treasure ships.

The target ship was the *Golden Gate,* a Pacific Mail Steamship Company boat that had gone down in 1862 off Manzanillo, on the West coast of Mexico. It was loaded with California gold, some $1,400,000 dollars worth. A dock would be built out from the shore, a coffer dam installed, the water pumped out, and heavy machinery used to dig into the wreck once it was located. Use of a heavy dredge to recover treasure from the shore would be another first.

By December 1979 the site had been identified by scuba divers and work was well underway. The expeditionaries again were camped on the beach, the site guarded by Mexican marines, and once again, as he had for the past twenty years CEDAM's veteran photographer, Genaro Hurtado, was making a film on the project.

When this excavation was finished, would CEDAM over the years ahead, with the backing of the Mexican government, probe the Mexican Caribbean for sunken gold? Or would it, and CEDAM International, do not only this, but also all the other things beckoning like a siren from the sea?

Time would tell.

Note

In March 1980, CEDAM International held "A Three Nation Archaeological Symposium" (Mexico, Canada, and the U.S.) in El Paso, Texas, cosponsored by the Center for Continuing Education, University of Texas at El Paso.

Members and guests were welcomed by the mayor of El Paso, the Hon. Thomas D. Westfall. Lectures included: Dr. Donald Rathbun, "Meteorites on the Bottom of the Sea"; Story Musgrave, NASA space scientist, "Scuba Diving Training for Astronauts"; Jack B. Irion, nautical archaeologist, "The Forty-Cannon Wreck Reveals Her Secrets"; Roger C. Smith, nautical archaeologist, "Grand Cayman Islands—A Graveyard of Sunken Ships," and Richard W. Underwood, NASA space scientist, "Photographs of the Earth Taken from a Manned Space Craft over Yucatan and the Central America area."

CEDAM founder Don Pablo Bush Romero, who lives in El Paso, took visitors on a trip to the Ciudad Juarez Public Museum and Art Center across the border, as well as arranged a display of underwater artifacts from CEDAM's museum at Akumal, Mexico.

In May 1981, CEDAM International held its Fourteenth Annual Members Meeting in Chicago in association with the Eleventh Annual Our World Underwater Seminar and Film Festival. Visitors heard CEDAM Advisor Robert Marx speak on "Treasures of the Indian Ocean," Al Giddings on the "Great White Shark," and Dr. Sylvia Earl on "Lockout at 1,200 Feet in the Revolutionary Jim-Suit." Winning underwater films were screened at the famous Chicago Civic Opera House.

In both 1980 and 1981 CEDAM International president

Bernarr MacFadden organized expeditions to Belize, Central America, not far south of CEDAM's old haunts in the Mexican Caribbean. Members of CEDAM International and CEDAM of Mexico met at the Paradise Hotel, Ambergris Cay.

Dick Chamberlain and Earle Crum were co-leaders, taking the group to Laughing Bird Cay as base of operations. From here the search widened to North Spot Cay where in a few minutes they located a wreck site with two coral-encrusted cannon.

They called this the "North Spot" wreck. Jack Irion, CEDAM International nautical archaeologist, reminded members they were custodians of the wreck and guided them in a professional search each year. Many artifacts have been uncovered. Belize's Minister of Archaeology, Harriot Topsey, gave permission to bring the pottery fragments back to the United States for evaluation.

Those interested in joining CEDAM International and participating in future meetings and expeditions should write: CEDAM International, P.O. Box 224046, Dallas, Texas 75264. This is a nonprofit organization with dues, fees, and contributions tax deductible.

Who's Who in CEDAM International

Many of the leading figures concerned with underwater exploration, marine research, and terrestrial archaeology are to be found within the ranks of CEDAM International.

OFFICERS

MICHAEL J. KELLY	Chairman of the Board
BERNARR MACFADDEN	President
THOMAS B. FIFIELD	Secretary
ROBERT S. GRUHN	Vice President, General Counsel
JOE KELLY HUGHES	Vice President, Operations
ELYSE L. MACFADDEN	Treasurer & Assistant Secretary

BOARD OF DIRECTORS

TAYLOR CAFFREY
RICHARD CHAMBERLAIN
CHUCK W. ENNIS
ALFONSO ROMERO ERAZO
THOMAS B. FIFIELD
ADM. HOWARD GREER
NIXON GRIFFIS
JOE KELLY HUGHES
MICHAEL J. KELLY
AMB. JAMES C. LANGLEY
BERNARR MACFADDEN
DR. PEDRO A. ORTEGA
PAUL H. BUSH ROMERO
ROMAN RIVERA TORRES
MICHAEL L. WILKIE
DOROTHY F. WINNETTE

BOARD OF ADVISORS

DEWEY L. BERGMAN
RALPH S. BUTLER
DR. SOL HEINEMANN
KENNETH R. HOLLINGSHEAD
WILLIAM P. HUGHES
ROBERT O. LEE
JOHN C. MAHON
M. TIMOTHY O'KEEFE
DR. CHARLES V. PEERY
DR. FRED SATTLER
EARL J. WILSON

HONORARY ADVISORS

ALFONSO ARNOLD
DR. GEORGE F. BASS
DR. JUNIUS BIRD
GERTRUDE DUBY BLOM
DR. ROLLIN H. BAKER
VICE ADM. A. CANIZARES
M. SCOTT CARPENTER
DR. GEORGE F. CARTER
DR. EUGENIE CLARK
GEORGE M. CLARK
GEORGE COE
CARL B. COMPTON
GORDON ECKHOLM
DR. HAROLD EDGERTON
WALTER FEINBERG
PETER R. GIMBEL
HAROLD L. GOODWIN
GENARO HURTADO
KENNETH W. McLAREN
DR. WILLARD F. LIBBY
JON LINDBERGH
EDWIN A. LINK
LUIS MARDEN
SIR ROBERT F. MARX
MELVIN M. PAYNE
MENDEL L. PETERSON
COLES PHINIZY
TAYLOR A. PRYOR
DR. STUART STRUEVER
PETER THROCKMORTON
ALBERT A. TILLMAN
TEDDY TUCKER
PAUL J. TZIMOULIS

COUNCIL OF EAGLES

WILLIAM F. CRAWFORD
THOMAS B. FIFIELD
NIXON GRIFFIS
ROBERT C. HANKEN
THOMAS Z. HAYWARD
MICHAEL J. KELLY
BERNARR MACFADDEN
DAVID KENNETH MORGAN
LARRY J. NERI
PAUL H. BUSH ROMERO
MRS. HAROLD SHEILA SCHAFER
PIERRE ELLIOTT TRUDEAU
MICHAEL L. WILKIE

Index